EINSTEIN'S LUCK

John Waller was born in England in 1972. He read
Modern History at the University of Oxford and went
on to take Masters degrees in Human Biology and in
the History of Science and Medicine. His doctoral
research was carried out at University College London,
where he is now a Research Fellow at the Wellcome
Trust Centre for the History of Medicine. He has taught
at the universities of Harvard, Oxford, and London.
Dr Waller is the author of *The Discovery of the Germ:
Twenty Years That Transformed Our Understanding of Disease*
(2002). He lives with his wife in Melbourne.

Read more about John Waller and his writing at
http://www.johnwaller.co.uk/

EINSTEIN'S LUCK
The truth behind some of the greatest scientific discoveries

John Waller

OXFORD
UNIVERSITY PRESS

OXFORD
UNIVERSITY PRESS

Great Clarendon Street, Oxford OX2 6DP

Oxford University Press is a department of the University of Oxford.
It furthers the University's objective of excellence in research, scholarship,
and education by publishing worldwide in

Oxford New York

Auckland Bangkok Buenos Aires Cape Town Chennai
Dar es Salaam Delhi Hong Kong Istanbul Karachi Kolkata
Kuala Lumpur Madrid Melbourne Mexico City Mumbai Nairobi
São Paulo Shanghai Taipei Tokyo Toronto

Oxford is a registered trade mark of Oxford University Press
in the UK and in certain other countries

Published in the United States
by Oxford University Press Inc., New York

First published 2002

First published as an Oxford University Press paperback 2004

British Library Cataloguing in Publication Data
Data available

Library of Congress Cataloging in Publication Data
Data available

ISBN 0–19–860939–6 (UK) *Fabulous Science*

ISBN 0–19–280567–3 (US only) *Einstein's Luck*

1

Typeset by Footnote Graphics, Warminster, Wilts
Printed in Great Britain
on acid-free paper by Clays

To my parents, with all my love

CONTENTS

ILLUSTRATIONS

ACKNOWLEDGEMENTS

Each chapter in this book draws heavily on the original scholarship of leading historians of science. The scholarly papers I consulted are the results of thousands of hours of laborious research, conducted by historians who often began their investigations confident of the veracity of the myths they would later explode. To pay these researchers their due, and to suggest where the reader may go to read further, I have included a brief statement on my sources at the end of the volume. In many instances I have added my own analysis and commentary, so any resulting errors or infelicities must of necessity be my responsibility alone.

This book is also indebted to the scholars who have stimulated my fascination for the discipline of history from school through university: Thomas Eason, Jeffrey Grenfell-Hill, Simon Skinner, Maurice Keen, the late Michael Mahoney, Joe Cain, Janet Browne, and Everett Mendelsohn in particular. The students who I have helped instruct at Oxford, London, and Harvard universities inspired me to bring these stories to a wider audience. And the staffs of Imperial College London's Centre for the History of Science, Technology and Medicine and London's Wellcome Trust Centre for the History of Medicine drew my attention to much of the scholarship upon which this book is based. The archival staffs of the Wellcome Library for the History and Understanding of Medicine, Harvard's Baker Business Library and Countway Medical Library, the Alexander Fleming Laboratory Museum at St Mary's Hospital, the Royal Astronomical Society, and the California Institute of Technology very kindly allowed me to reproduce photographic images.

I am also profoundly grateful to Richard and Penny Graham-Yooll, Adam Hedgecoe, Darian and Alison Stibbe, Jon Turney, and Jane, Michael, and Susan Waller for their excellent editorial advice and encouragement. My father, Michael Waller, has been a wonderful source of inspiration, illumination, and guidance. Michael Rodgers and Abbie Headon at Oxford University Press and copyeditor Sarah Bunney have also been most helpful. Finally I would like to thank my wife, Abigail, for her unstinting affection and support.   *John Waller, December 2001*

INTRODUCTION
WHAT IS HISTORY FOR?

Remark all these roughnesses, pimples, warts, and everything
as you see me, otherwise I will never pay a farthing for it.

> Oliver Cromwell's instruction to Lely
> on the painting of his portrait.

The great biologist Louis Pasteur suppressed 'awkward' data because it didn't support the case he was making. Gregor Mendel, the supposed 'founder of genetics', was no Mendelian. Joseph Lister's famously clean hospital wards were actually notoriously dirty. Alexander Fleming misled the world about his role in the discovery of penicillin. And Einstein's general relativity was only 'confirmed' in 1919 because an eminent British scientist ruthlessly massaged his figures.

These are some of the recent findings of historians covered in this book. In writing it my primary aim has been to bring to the attention of a wider audience the fruits of a generation's research into the history of science. But, although the scholarship I draw on seriously challenges the reputations of major scientific figures, this is not an exercise in pointless iconoclasm. Above all, this book aims to offer insights into the conduct of scientific debate, the securing of scientific immortality, and the complex interplay between scientists and the worlds in which they operate.

In highlighting these 'warts and all' studies, I am firmly positioning myself on one side of a great historical divide. As the divide in question runs through the entire historical enterprise, I can illustrate it with an example taken from Classical Rome. These are the opening lines from 'Horatius', the first in Thomas Babington Macaulay's epic series of poems, *Lays of Ancient Rome*, written in 1842:

> Lars Porsena of Clusium
>   By the Nine Gods he swore
> That the great house of Tarquin
>   Should suffer wrong no more.
> By the Nine Gods he swore it,
>   And named a trysting day,

Previous page: The 'dauntless three' guard the bridge over the Tiber, from Macaulay's epic poem *Lays of Ancient Rome*.

> And bade his messengers ride forth,
> East and west and south and north,
>   To summon his array.

A favourite of the Victorian schoolroom, this poem still has the power to quicken the pulse and enthuse the reader with martial pride. What seems to make the *Lays of Ancient Rome* doubly stirring is its historical veracity. Macaulay drew his great poem not from his admittedly fertile imagination, but from the Roman histories themselves. These tell how in the sixth century BC the enemies of Rome led by the Etruscan Lars Porsena had fought, sacked, and plundered their way to the shores of the Tiber. At last, their greatest prize—the eternal city itself—lay all but defenceless before them. The battle-hardened troops of a long and glorious campaign gazed across to a city facing slaughter, rapine, and ruin. But a quick-thinking Roman Consul had seen a way to save his people. The plan was simple. The only bridge over the Tiber was barely wide enough for one man to pass at a time. So a handful of men could hold the bridge long enough for it to be cut away behind them. As these valiant soldiers plunged to a watery death, they would have the consolation of fulfilling every Roman matron's deepest desire for her son: his giving his life to save the Nation. This notwithstanding, so tough was the challenge that at first only brave Horatio accepted it. Then two others followed his example.

It was this 'dauntless three' that confronted the Etruscan host as it approached the bridge. As the Consul had foreseen, the narrowness of the structure meant that one-to-one fighting was all that was practical. The odds thus reduced, Horatio and his colleagues slew the ablest champions Lars Porsena could throw at them. As the bridge started to fall, Horatio's two companions made it back to shore as he covered their retreat. For him there was no escape. Valiantly out in front, he could do nothing except prepare to meet his end. Exhausted, wounded, and weighed down with armour, Horatio plunged beneath the surface of the Tiber. A stunned calm fell upon Rome and her enemies:

> No sound of joy or sorrow
>   Was heard from either bank;
> But friends and foes in dumb surprise,
> With parted lips and straining eyes,

Stood gazing where he sank;
And when above the surges,
They saw his crest appear,
All Rome sent forth a rapturous cry,
And even the ranks of Tuscany
Could scarce forbear to cheer.

Horatio lives! Swimming back to the Roman shore, his feat of daring and courage momentarily united the warring sides in admiration and awe. Who knows how long it was before the invaders' sentimental feelings gave way to bitterness? But what is clear is that Horatio's courage saved Rome. In recognition of this, his fellow citizens lavished tributes and prizes on him. Then, again according to Macaulay, to immortalize the event, 'they made a molten image, / And set it up on high, / And there it stands unto this day / To witness if I lie.'

Look around the ruins of Rome today, however, and you will search in vain for the statue of Horatio. Of course, given the amount of destruction that has occurred there, this does not necessarily give the lie to the story. But it *is* well and truly laid to rest by the detailed and reliable chronicles of the sixth century BC, which describe how the Etruscan armies swept irresistibly upon Rome and overwhelmed it. Within hours, those inhabitants of Rome who survived the onslaught would have been forming the long, miserable columns typical of refugees in flight. These chronicles also show that the only laurels earned during the campaign decorated the brow of Rome's most implacable enemy, Lars Porsena. Of Horatio, there is nothing. Rome's humiliation, it is clear, was sudden, swift, and far from painless.

So how did this fairy tale manage to get airborne? Partly because by the time the Horatio myth took hold Rome had become the world's mightiest empire and she needed to invent a past of sufficient grandeur to justify and glorify her present. The myth of Horatio did nicely because it seemed to show that the Romans had always been great warriors, brave fighters, and brimming with guile. But the popularity of the Horatio story goes even deeper than this. By the first century AD the Romans were too proud and too intoxicated by their new wealth and power to be capable of believing that Lars Porsena had sacked Rome. In this heady atmosphere, myth became established fact. The Horatio story, in other words, reflects not history, but how a later generation of Romans needed to see their forebears.

Robert Graves's *I, Claudius* includes a wonderful scene in which the Roman historians Livy and Pollio use the Horatio story to debate precisely the two rival approaches to history I am trying to illustrate. Their exchange is unintentionally brought about by the stammering young Claudius. Having warmly praised Livy's style, he goes on to say that he is somewhat puzzled by inconsistencies between Livy's version of the Etruscan War and evidence he has uncarthed that, in fact, Rome was defeated. Although Livy seeks to dismiss the counter-evidence as mere propaganda, he soon makes clear that in his opinion the role of history is not the uncovering of the Truth. Instead, it is that of arresting moral decline by providing idealized role models to the young and of dignifying the present through association with a glorious past. In contrast, Pollio secures Claudius's support for the rival view that the historian's ultimate duty is to the Truth.

There can be no doubt at all that, until very recently, Livy's view of history triumphed over Pollio's as convincingly as did Lars Porsena over the Romans. About 1700 years after Livy's death, Macaulay happily recycled Livy's *Ab Urbe Condita I* as the *Lays of Ancient Rome*. The imperial power had changed, but the inspirational objectives remained the same. Among subsequent generations of classical historians (albeit posthumously), however, Macaulay met his Pollio. Wonderful though it is as a poem, the story told in the *Lays of Ancient Rome* is now firmly consigned to history's dustbin. Nor is it alone. Acutely conscious of the human predilection to glamorize, embellish, and invent, historians over the past hundred years have turned the same penetrating and unromantic gaze on all fields of human endeavour. Pollio's view of the proper role of history is now in the ascendant.

But establishing precisely where the truth lies is rarely easy. Personal biases of which the researcher may be unaware, a lack of unambiguous data, and the difficulty of seeing the world as our ancestors saw it can sometimes combine to make the task virtually impossible. Yet a neo-imperialist in search of a modern Livy would be hard-pressed to find any professional historian still comfortably working in this tradition. If 'professional' in the phrase 'professional historian' means anything, it means an unambiguous commitment to seeking out the truth. Execution may not always match aspiration, but that does not significantly diminish the scale of the change that has taken place. To anyone who resents the loss of his or her particular Horatio there is but one answer. Now, at least, you are being told what can best be assayed as the truth.

Telling science as it is

When first thought about, it might seem that science is a field in which propaganda of the type favoured by Livy would have little or no scope. After all, science itself is all about the search for truth and the adversarial nature of the scientific method should work relentlessly to ensure that only valid ideas supported by well-designed experiments survive. I think that after reading this book, the most likely response to such notions will be 'but that they were so'. Because they have been sought out to demonstrate the gap that can exist between myth and reality, the various case studies I include are not offered as representative of all science or all scientists. But one general message of the utmost importance can be drawn from them. Even in the realms of science, take nothing at face value.

Until recent decades, the history of science was largely written by those who wished to place their chosen subject in as favourable a light as possible. Their motivations were various. Sometimes they worked at the behest of individual scientists who wanted to make sure that their part in the great drama of discovery did not go unsung. In other cases, the key requirement was a good story. More laudably generations of teachers of scientific subjects have wanted heroes for much the same reason that Livy gave the Romans Horatio: to inspire by example. The chosen ones entered the Pantheon of scientific heroes. Great laboratories and institutes were named in their honour; each new generation of students was given accounts of their travails and ultimate triumphs; and assorted statuary serves as a perpetual memorial to their greatness.

In the last few decades, however, this approach has been rightfully impugned. A new generation of scholars has shown that in many cases what actually happened simply cannot sustain the enormous edifice subsequently built on it. Many of the great luminaries of the past were neither as heroic nor selfless as has been supposed. Seemingly crucial experiments are sometimes found to have been fatally flawed; results were often modified to suit the case being argued; and many were happy to use political influence to advance their cause. Indeed, ample evidence is now available to show that scientific merit is only one of many factors influencing the acceptance of new ideas. Many pre-eminent scientific heroes fell far short of proving the theories for which they are now famous. Men such as Louis Pasteur, Joseph Lister, and Alexander Fleming were neither as sure-footed nor as scrupulous as they are now thought to have been. Charles Darwin was right at least partly for the wrong reasons. Others, such as Gregor Mendel, have had greatness thrust on them by a highly manipulative posterity. And, not infrequently, individuals now cast as scientific villains prove on closer examination to have been able scientists who just happened, often for very good reasons, to have backed the wrong horse.

Above all, what this new research shows is that the conduct of scientific enquiry is often a lot more haphazard than we tend to think. Although the eventual outcome of a research programme may be a fabulously rich collection of well-attested and highly predictive ideas, the route to this happy state is often far more convoluted than subsequent accounts will allow. Revealing what actually happened in some very high-profile cases may help bring our conception of the scientific enterprise into much closer alignment with the actuality. None of this undercuts the status I believe modern science deservedly enjoys as the best way of increasing our understanding of the physical world. But our expectations will be more realistically grounded if we come to appreciate that science is as subject to extraneous influences—including the human ego—as is any other field of human endeavour, past or present.

There is another important service that historians of science can render. As in all other branches of history, 'great man' approaches massively underplay the contributions made by the myriad individuals who did not achieve this honoured status. Thousands upon thousands of now largely forgotten researchers have contributed to scientific progress. And with very few exceptions, great men or women are cumulatively far less important than these forgotten legions of unsung heroes about whom little is popularly known. Indeed, some mute inglorious scientists were just as insightful and technically ingenious as those whose names have lived on. In many such cases, the differences in historical treatment are best explained in terms of a general preference for attaching major ideas to a limited number of names, coupled with skills, or the want of them, in the arts of self-promotion. The pristine hero, exemplified by brave Horatio, is all too often an elaborate fiction. If we go back and look at the primary sources, few reputations escape entirely unscathed.

In the context of these broader considerations, I have tried to use these case studies of nineteenth- and twentieth-century science to make three basic points. First, that we need to treat received accounts of scientific genius with the utmost circumspection. Thus we will find that Louis Pasteur, Charles Darwin, Gregor Mendel, Thomas Huxley, Joseph Lister, John Snow, Alexander Fleming, Frederick Winslow Taylor, James Young Simpson, Charles Best, Arthur Eddington, the Nobel Prize-winning Robert Millikan, and the authors of the famous 'Hawthorne Study' (Fritz J. Roethlisberger and William J. Dickson) have all been squeezed, or have squeezed themselves, into romantic schemas strongly redolent of the Horatio myth. In many cases, these men were competing for laurels in a highly competitive world in which the Queensbury rules of the scientific method were routinely dropped in favour of the more-permissive code of bare-knuckle fighting. The chapters I have devoted to Pasteur, Lister, Taylor, Millikan, Eddington, Best, and Roethlisberger and Dickson illustrate this particularly strongly. Indubitably, each of these scientific greats carried their share of human frailties.

The second point I seek to make is the critical importance of contextualization. Science is about much more than disembodied ideas. In each chapter I set the events described in the broader context essential to a full appreciation of the complexities involved in the process of scientific discovery. This book stresses the role of the prevailing scientific paradigm, the social and political context, and the vagaries of chance, all of which powerfully influence the rate and direction of scientific progress. Traditional approaches rarely accorded such factors their full weight. My hope is that these cases will demonstrate the critical importance of remedying this.

This allusion to context brings me to the final theme of the book. Failure to take full account of context leads to an error that modern historians call 'presentism'. Individual chapters dealing with Lister, Mendel, Darwin, Snow, Huxley, Simpson, and Fleming serve to elucidate this problem. A vague affinity between a currently accepted theory and a much earlier set of ideas is often enough to elevate the ancestor into the Pantheon of scientific heroes. Just as Macaulay, Margaret Thatcher, and very many others, have mistakenly read Magna Carta as an early flowering of English democratic values, those recounting the history of science have often wrenched older ideas entirely out of context and interpreted them as brilliant anticipations of modern knowledge.

Some of the greatest icons of science have acquired hero status in precisely this way. Put back into the context in which the originators lived out their lives, many ideas are found to be much less clearly aligned with what we now believe to be true. But we are taught to demand much of our founding fathers. Their having been there at the beginning, pointing the way forward, does not seem to be enough. There is also a tendency to expect them, long after they have entered the grave, to remain in the van of progress, their ideas at least broadly anticipating each new development. What we need to bear in mind is that the past really is another country and most certainly not one of which the present was an inevitable culmination. Therefore my third aim is to encourage contextualization not only in its own right but also as a sovereign remedy to presentism. We need to be committed to understanding the past on its own terms without any reference to 'what happened next'.

My final hope is that the cases I have chosen will prove fascinating in themselves as powerful human dramas in which naked ambition has at least as big a role as technical virtuosity. As such I am confident that they will prove particularly appealing to the group who it has been suggested are most willing to look at the world from the anti-presentist perspective: those who have a strong spirit of adventure coupled with a deep respect for the human intellect. To such minds, teasing out truth from fiction serves only to enrich their understanding of the human condition.

Louis Pasteur

Robert Millikan

Arthur Eddington

Frederick Winslow Taylor

Fritz J. Roethlisberger and William J. Dickson

The first five chapters share one common theme: each of the six major scientists examined manipulated their experimental data to fit their preconceived notions of how things really are. Then, to win the scientific battles in which they were engaged, they exploited (to varying degrees) their powers of obfuscation and deception, their friends in high places, and their reputations as reliable witnesses. All six have been fortunate in the fact that because they were advancing major ideas that now enjoy, at the very least, widespread support, posterity has been largely blind to the equivocal nature of the evidence they presented.

It would be unfortunate, however, were the next five chapters read as attacks on the scientific enterprise. Naturally showing how scientific debates can be distorted by historical context and the human ego does detract from science's reputation for unalloyed objectivity. Likewise, there is no avoiding the conclusion that some of the greats of the history of science sometimes let ambition get in the way of integrity and good science. But the six scientists I examine in the following chapters are not necessarily representative of science in general. I have selected them because of the gulf that separates the myths surrounding their names from the actuality. How much light they shed on scientific endeavour as a whole will be considered in the Conclusion to Part 1.

With respect to Pasteur, Millikan, and Eddington, at this stage I should like to make one further observation. The twentieth-century philosopher of science Karl Popper made a useful distinction between *discovery* and *verification* in the development of scientific knowledge. A committed and eloquent believer in the ability of scientists to make sense of the world, he nonetheless saw that the discovery stage may be much less rigorous and disciplined than the point at which other scientists become involved and begin the process of verification by trying to 'falsify' the original researcher's ideas. 'The question how it happens that a new idea occurs to a man—whether it is a musical theme, or a dramatic conflict, or a scientific theory—may be of great interest to empirical psychology; but it is irrelevant to the logical analysis of scientific knowledge' is how Popper put it in 1959. Science only becomes reliable knowledge, he argued, after its validity has been extensively tested over the course of many years. Indeed, he was 'inclined to think that scientific discovery is impossible without faith in ideas which are of a purely speculative kind, and sometimes even quite hazy'.

This is the context in which the cases looked at in Part 1 need to be understood. The initial evidence presented by these scientists was seriously flawed and they were each led more by conviction than empirical data. But had they been unequivocally wrong about the way in which the world operates, then research by other scientists in other laboratories would soon have shown this to be the case. Incorrect but plausible ideas have often been endorsed by sections of the modern scientific community. Almost never, however, have they stuck around for long: a theory must have considerable merits for it to stand a chance of survival in a milieu that thrives on disagreement.

That said, I think there is plenty in the next few chapters that the reader will find eye-opening. This is because the true complexities of Popper's discovery stage are not widely appreciated; even Popper had very limited opportunities for effectively researching them. The cases looked at here are important, therefore, in that they indicate just how tricky, uncertain, and byzantine a business scientific discovery can actually be. Contrary to the traditional view, this critical stage is mediated by a wide range of social and psychological factors that all too easily tempt researchers from the path of righteousness laid down by the rules of the scientific method as conventionally defined. During the verification stage, the prognosis for bad ideas supported by good PR is extremely poor. But when new territory is being opened up, there is far more scope for tactical skills and sheer force of personality to play decisive parts. It is to five such cases that I now turn.

1

THE PASTEURIZATION OF SPONTANEOUS GENERATION

Here was a life, within the limits of humanity, well-nigh perfect. He worked incessantly: he went through poverty, bereavement, ill–health, opposition: he lived to see his doctrines current all over the world, his facts enthroned, his methods applied to a thousand affairs of manufacture and agriculture, his science put in practice by all doctors and surgeons, his name praised and blessed by mankind . . . Genius: that is the only word . . . In brief nothing is too good to say of him.

Stephen Paget (British scientist), *The Spectator* (1910).

Pasteur's recognition of the fact that both lactic and alcohol fermentations were hastened by exposure to air led him to wonder whether his invisible organisms were always present in the atmosphere or whether they were spontaneously generated. By means of simple and precise experiments, including the filtration of air and the exposure of unfermented liquids to the air of the high Alps, he proved that food decomposes when placed in contact with germs present in the air, which cause its putrefaction, and that it does not undergo transformation or putrefy in such a way as to spontaneously generate new organisms within itself.

'Louis Pasteur', *Encyclopaedia Britannica* (1992).

In 1878, as Louis Pasteur's fame in scientific circles was approaching its zenith, his venerated friend Claude Bernard collapsed in his laboratory at the Collège de France in Paris. Not even Bernard, the recognized master of nineteenth-century physiology, could arrest the kidney infection that would kill him a few days later. To the germ responsible, Pasteur believed that he had lost both a friend and one of his most constant

Left: Louis Pasteur (1822–95) in his laboratory.

supporters in the highly competitive scientific milieu of Third Republic France.

But Pasteur was in for a rude shock. After 6 months had passed, one of Bernard's student admirers published portions of his deceased mentor's laboratory notebooks. Pasteur was horrified, for these sketchy notes contained remarks explicitly prejudicial to his own scientific work, written by a man who still ranks among the world's most-celebrated practitioners of experimental science. Pasteur, Bernard's notes claimed, held to the germ theory of disease more on the basis of preconceptions than scientific evidence. Appalled, Pasteur rushed into print. Systematically refuting Bernard's criticisms, he loosed a broadside that many of the latter's acolytes considered to be an act of inexcusable desecration. In an unashamed tit-for-tat vein, Pasteur alleged that it was Bernard himself who had fallen victim to the 'greatest derangement of the mind': the 'tyranny of preconceived ideas'.

To those who knew him well, Pasteur's vindictiveness came as no surprise. Prodigiously clever, exceptionally hard working, and a superlative organizer, Pasteur was also intensely ambitious and very touchy about criticism. This far from unusual combination of prickliness and high aspirations meant that in proving his point he could be both insensitive and ruthless. Pasteur's intolerance of lax thinking once even provoked an 80-year old adversary into challenging him to a duel (luckily it did not take place). His character also led him to expect unconditional loyalty from his assistants and underlings. For the most part he probably deserved it. But Pasteur was not averse to advancing their ideas as his own and swearing them to secrecy when he did so. Most went to their graves knowing that their work had been appropriated by their mentor. Indeed, Pasteur thought of his laboratory and all that went on within it much as Louis XIV had seen the French state, as a personification of himself. The downside of this was that few accomplished scientists were willing to work under him: Pasteur found protégés very hard to come by. Nevertheless, for those willing to swallow their pride, there were definite benefits to be derived from working under someone with the self-belief to reassure his wife at the age of thirty that he would 'lead her to posterity'. With a bit of luck—and Pasteur had more than his fair share—such people are a rich source of reflected glory.

In Pasteur's dispute with Bernard's ghost, philosophy merged with psychology. The two accused each other of failing to meet one of the cardinal principles of the scientific method. The experimental phase must rigorously test the theory; it must not itself be shaped or modified by an imperative to prove the theory correct. To accuse somebody of being tyrannized by 'preconceived ideas' is to suggest that just such a process of shaping and modification has occurred. Implicit in this accusation is the claim that the experimenter has stood between his or her findings and the rest of the scientific community, filtering out any results and eliminating any methods that seem likely to discomfit their own view.

To be shown to have committed such a transgression would have seriously damaged Pasteur's career. Fortunately for him, with Bernard not around to press his case, the savage counter-attack proved successful. By the time of his own death in 1895, Pasteur had won for himself an international reputation as France's premier scientist. Celebrated throughout the world for championing the germ theory of disease and for tirelessly promoting the practices of vaccination and heat sterilization ('pasteurization'), he enjoyed fame on a scale now largely reserved for sportsmen and film stars. As the obituary written by Stephen Paget quoted above suggests, this was a scientist with cult status.

Pasteur won much of his immense prestige during the 1860s when he consigned the concept of 'spontaneous generation' to the scrap heap of discredited scientific theories. With a series of experimental set pieces that have become classics of the history of science he won a highly dramatic and public dispute. In this and other exhibitions of technical skill, Pasteur showed that he was an expert practitioner of the experimental method. Accordingly, in the days following his death the scientific community enthusiastically added his name to the growing Pantheon of first-class heroes in the history of science. Since then, a century's worth of biographies have consistently cast him as the scientists' scientist. As such, Pasteur is viewed as a man of absolute integrity whose work gloriously embodied the prejudice-free nature of experimental science. When coupled with the exceptional quality of his ideas, this exemplary approach served to vindicate his own theories and reveal those of his opponents to be little more than superstitious nonsense. Those holding this view accord his opponents no more than walk-on parts in which they are ritually humiliated by a superior show of logic and experimental virtuosity. Having been bested by the great Pasteur they quit the field, either piqued at their misfortune or ready to pay homage to the victor.

Yet, although Pasteur's is a truly inspirational story, as the historians Gerald Geison and John Farley have shown, many aspects of it are seriously at variance with the facts. Drawing on Geison's and Farley's re-appraisals of Pasteur's refutation of the theory of spontaneous generation, this chapter reveals a man who certainly did not live up to the exacting standards of scientific practice he admonished Claude Bernard for forsaking. By drawing out critical elements of the context in which the debate over the origins of life took place, we see that Pasteur was definitely not unprejudiced and that his experimental evidence on the role of germs in putrefaction and fermentation was a long way from being decisive. Ulti-mately it fell to the more sophisticated research of later German scientists to prove conclusively that Pasteur had been in the right.

'Life is the germ, and the germ is life'

To believe in spontaneous generation was to think that primitive life forms can arise without the involvement of either parent organisms or supernatural forces. Its advocates argued that the micro-organisms observ-able in putrefacting matter are produced *in situ*; rather than causing decay, germs arise from the sudden creation of entirely new life in decaying matter. Instead of using modern knowledge to ridicule this idea, we need to recognize how gradually the evidence that underpins the germ theory of disease accumulated during the nineteenth century. We also need to see that by the 1860s, spontaneous generation's foremost French pro-ponent, the elderly Rouen naturalist Felix Pouchet, had amassed a great deal of what seemed to be hard evidence supporting the idea. And Pouchet was certainly no crank. During the 1840s he had shown that, contrary to prevailing wisdom, ovulation is not activated by male sperm, a finding that ought to have earned him an honourable place in the scient-ific hall of fame. Instead, posterity has cast him as Hotspur to Pasteur's Prince Hal.

The prolonged battle between Pasteur and Pouchet was initiated in 1858 when Pouchet circulated a paper at the Académie des Sciences—the nerve centre of French science—claiming that he could produce an experimental vindication of spontaneous generation. He explained how he had heat-sterilized a quantity of hay, exposed it to artificially produced air or oxygen, and separated it from atmospheric air with mercury. In this presumably sterile hay infusion, Pouchet claimed to have detected the remarkable—*de novo*—appearance of micro-organisms. This claim directly challenged the now-accepted counter-view that airborne micro-organisms themselves cause putrefaction and that in a sterile atmosphere decomposition cannot take place. The stage for a grand controversy was set.

Two years after Pouchet circulated his paper, Pasteur, who had established his name in crystallography, made public his intention of accepting the challenge. It was precisely what he had been looking for: serious scientific research with enormous public appeal. He was about to start his journey from private to public scientist and to deliver the immortality he had promised his wife several years earlier. It was to be worth her wait. Four years of experimentation later, on 7 April 1864, Pasteur took his position on the stage at the Sorbonne's packed amphitheatre and outlined a series of experiments he had devised and carried out that seemed to prove Pouchet's claims false. Only the confrontation between Thomas Huxley and Bishop Wilberforce in Oxford several years before (Chapter 10) comes close to this event for dramatic and symbolic effect.

Standing before the cream of French political and intellectual society, Pasteur began by explaining how, in 1860, he had trapped the solid contents of atmospheric air in a piece of guncotton. This 'atmospheric dust' was then treated so that it could be examined under a microscope. Although his method inevitably killed any micro-organisms that were present, the 'corpuscles' seen by Pasteur through the eyepiece looked unmistakably like the remains of living organisms. Next, he had attempted to demonstrate that micro-organisms do not appear in a heat-sterilized solution unless that solution is subsequently exposed to atmospheric air.

Pasteur's first method involved replicating Pouchet's experiment with a sterilized organic solution in a mercury-filled trough. Not entirely satisfied with this method, Pasteur devised a second apparatus. Into a flask he poured a quantity of sugared yeast-water that he boiled for several minutes. Air that had first been sterilized by being passed through a red-hot platinum tube was then introduced to fill the airspace above the yeast solution. Following this, the flask was sealed with a flame and placed in a stove held at a temperature known to be conducive to microbial growth. After 6 weeks Pasteur removed the flask from the stove and, having noted that there was no evidence of life, inserted a small wad of guncotton charged with 'atmospheric dust' without permitting the entry of atmospheric air. After 24–36 hours the once-sterile fluid was thick with micro-organisms.

Anticipating the accusation that micro-organisms had been generated spontaneously from organic material in the guncotton, Pasteur repeated the experiment using asbestos, the mineral, in its place. Again, the atmospheric dust, this time introduced on the asbestos, brought about the emergence of micro-organisms in what immediately before had been a sterile solution. Thus, Pasteur announced, he had proved conclusively that micro-organisms appeared only when the fluid had been contaminated with the solid particles of atmospheric air.

Subsequently Pasteur further refined his methods by use of his famous 'swan-necked' flasks. Rather than applying heat to seal these, he narrowed, attenuated, and contorted their necks to such a degree that, when stored in a still room, atmospheric air did not interact with the flasks' contents. The sugared yeast-water in most of the flasks was heated, although a few were left untreated as controls. After storage for 24–36 hours, the unboiled liquids were covered with mould whereas the boiled flasks remained unaltered. Deprived of contact with atmospheric particles, the solutions remained sterile. As an encore, Pasteur broke off the necks of the mould-free flasks and dropped them in the fluid mixtures. As he predicted, mould soon formed on their surfaces as the broken necks introduced atmospheric air into the flasks. Only 'germs' borne by atmospheric air, he again concluded, could possibly explain these empirical observations.

Although these ingenious experiments gave round one comfortably to Pasteur, the contest was far from over. Pioneers of food canning technology were just then claiming that putrefaction could occur if organic material was exposed to the tiniest amount of oxygen. This presented Pasteur with a difficult problem. He had argued that different forms of micro-organism cause all the different kinds of putrefaction and fermentation that were commonly observed. But this notion was difficult to square

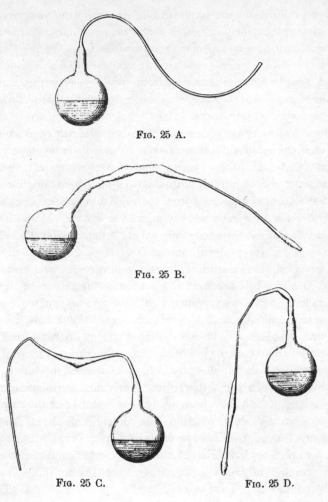

FIG. 25 A.

FIG. 25 B.

FIG. 25 C.   FIG. 25 D.

Pasteur's famous swan-necked flasks.

with the minute quantities of air that Pouchet and the canners claimed were all that was necessary for new life to emerge. How could so many different types of micro-organism be present in so small a volume of gas?

In November 1860, Pasteur's attempts to deal with this objection took him 2000 metres above sea level on the Mer de Glace glacier in the

Alps. Working on the assumption that the quantity of micro-organisms in air varies in accordance with the density of organic matter in the immediate environment, he had spent the previous weeks exposing pre-sterilized flasks of boiled, sugared yeast-water at various altitudes and locations: those carried up to the glacier were the final set. As with the others, having been exposed the flasks were resealed and removed to a stove held at temperatures ordinarily conducive to the growth of micro-organisms. Confirming Pasteur's expectations, the less microbe-rich a flask's point of exposure, the less likely its contents were to undergo fermentation. High up on the Mer de Glace, oxygen alone had not been enough to induce fermentation. Pouchet received another well-publicized humiliation.

Still Pouchet and his supporters remained unrepentant. They simply responded that by overheating the sugary yeast-water solutions used on the Mer de Glace Pasteur had destroyed the 'vegetative forces' needed to create new life. In 1863, with outstanding technical skill, Pasteur therefore collected blood and urine directly from the veins and bladders of healthy cattle. These mediums did not require heating to be sterilized and, as in his previous experiments, micro-organisms appeared only on exposure to atmospheric air. A year later, in the hallowed ground of the Sorbonne amphitheatre, Pasteur delivered what many considered to be the *coup de grâce*. Returning to Pouchet's experimental procedure of using boiled hay infusions and mercury troughs, he showed that although his adversary had been careful to sterilize his organic material and use non-atmospheric air, he had not been nearly so scrupulous with the only other possible source of contamination—the mercury. Displaying his usual virtuosity, Pasteur provided experimental evidence strongly suggestive of Pouchet's mercury having have been left exposed to atmospheric air and having consequently been the source of microbial agents.

This final tour de force so invigorated the assembled ranks of French high society that at the close of the lecture they gave Pasteur a standing ovation. This is partly because high science and high culture were then far more intertwined than today. Particularly in France, self-respecting intellectuals made an effort to keep abreast of scientific findings. But there was also another factor working to Pasteur's advantage. The previous 60 years had witnessed the convulsions of the Terror, the bloody collapse of Napoleon I's imperialist ambitions, the restoration of the monarchy, its usurpation by the Bourbons, and now the ascendancy of Napoleon's nephew, Louis Napoleon. To many that evening it must have seemed that Pasteur's strictly regulated flasks and troughs represented a blessed haven of rationality. In the world outside, the resolution of argument without recourse to the militant *sans culottes*, the sword, the barricade, and the *coup d'état* seemed an illusory hope. There, 'truth' was the preserve of the highest bidder and the hardest hitter. Against this backdrop, Pasteur's experiments fed a profound yearning for the conclusive, the fair, and the disinterested—qualities so tragically absent from the spheres of politics and religion.

The net effect was that by the end of this pivotal lecture, Pasteur had brilliantly established a consensus view. Addressing his rapturous audience, he slipped into metaphysics. By depriving the sugary yeast-water of germs from the air, he explained, 'I have removed from it the only thing that it has not been given to man to produce . . . I have removed *life*, for life is the germ, and the germ is life'.

Were Pasteur's results decisive?

Given such an account, it is hardly surprising that Pasteur came to be seen as what amounts to a secular saint. Under the gaze of modern scholarship, however, it is a story that rapidly falls apart. Although Pasteur was indisputably on the side that won, we can show that during the 1860s and 1870s he was never able to advance incontrovertible arguments against spontaneous generation. Indeed, the best indication that Pasteur was driven by conviction rather than hard evidence—as Claude Bernard had claimed—is provided by his own laboratory notebooks.

For a man claiming to have investigated the question of spontaneous generation 'without preconceived ideas', Pasteur's memoirs on the subject contain some surprising anomalies. In 1861, his experiments with mercury and sugared yeast-water had shown evidence of the growth of micro-organisms prior to the solution's exposure to atmospheric air in more than 90 per cent of cases. In other words, more than 2 years before he recognized that mercury itself can contaminate the organic solution, one of his key experiments provided exceptionally strong evidence in Felix Pouchet's favour. This apparent doyen of the Scientific Method later explained that he had not published these data 'for the consequences it was necessary to draw from them were too grave for me not to make them irreproachable'. This was hardly giving spontaneous generation a fair hearing. In fact, throughout his feud with Pouchet, Pasteur described in his notebooks as 'successful' any experiment that seemed to disprove spontaneous generation and 'unsuccessful' any that violated his own private beliefs and experimental expectations.

Pasteur's claim that Pouchet's experiments were invalid because he had used contaminated mercury are in themselves fascinating examples of a lapse in scientific logic. The phenomenon of scientists rejecting counter-evidence on the basis that the experiment had been performed incorrectly has been dubbed 'experimenter's regress'; it is especially common in fields where new ideas are being supported by new, untried, and difficult-to-use experimental apparatus. Experimenters often have no real way of knowing whether a failure to replicate a result obtained by another team reflects experimental errors on their part or on the part of the original investigators. Indeed, both factors may be at work. Pasteur seems to have been oblivious to such difficulties. In breach of a canonical rule of the scientific method and with a circularity that would have been condemned by posterity had Pasteur not happened to be correct, wherever possible he used 'contaminated mercury' as the catchall for ruthlessly rejecting any evidence that Pouchet put forward. Virtually the sole criterion for accepting or rejecting his own experimental findings was equally straightforward: whether or not they supported the theoretical position he had adopted. It would be an insult to Pasteur's high intellect to excuse this behaviour on the grounds that the scientific discipline was then far more lax. His counter-attack on Bernard shows just how savagely critical he was when accusing others of just such behaviour.

Pasteur also displayed a cavalier attitude towards replicating his rivals' experiments. In 1864, Pouchet repeated one of Pasteur's most-spectacular experiments by exposing a sterile solution to atmospheric air at high altitude in the Pyrenees. But he did so with one crucial difference. Instead of sugared yeast-water, he used a boiled hay infusion. The result was that all of his flasks developed mould consistent with the claim that only oxygen is necessary for life to begin afresh. Delighted with these findings, Pouchet once more threw down the gauntlet to the Académie in Paris.

But Pasteur flatly refused either to repeat Pouchet's Pyrenean experiments or to consider the possibility that the use of hay infusions in place of yeast-water might have made a difference. In this case he could not blame the mercury as none had been used. Instead, he fired off a single, vague complaint about Pouchet using a file rather than pincers, and then simply refused to discuss the Pyrenean experiments any further. The rest of the debate proceeded as if Pouchet's greatest and most promising moment had never actually occurred. Clearly, science and self-belief are not always happy bedfellows.

What made such behaviour all the more unreasonable was that Pouchet was perfectly entitled to argue that Pasteur's failure to produce spontaneous generation in his flasks could not resolve the question of whether organisms could appear anew in *other* circumstances. Advocates of spontaneous generation needed to find only one emphatic example of the phenomenon to win the debate. Instead of accepting this, Pasteur chose to defy logic by asserting that because micro-organisms were not spontaneously generated in his flasks, they could not come into being anywhere. With so little then known about the nature and mechanics of life, Pouchet's position was actually near-impregnable. Indeed, the impossibility of proving a negative—that is, of showing that spontaneous generation cannot happen under any circumstances—means that this question must for ever lie open.

We can go further. By those not prepared to accept a supernatural explanation for the origins of life, the assumption has to be made that somewhere in the Universe, at some time in the past, on at least one occasion, spontaneous generation has actually occurred. The latter consideration was not a problem for Pasteur because of his Catholicism. Nonetheless, from a scientific standpoint, it was beholden on him either to expose a flaw in each and every success Pouchet claimed, or concede the argument. Pasteur seems to have been aware of this and that is why Pouchet's Pyrenean experiments caused him considerable anxiety. He simply could not deny to himself the possibility that by altering the experimental conditions, Pouchet had been vindicated. The problem was that by 1864, Pasteur's emotional and professional investment against spontaneous generation was so great as to heavily outweigh his commitment to his own definition of the scientific method.

That this is the case is strongly suggested by Pasteur's shabby treatment of the University College London physiologist, H. Charlton Bastian. In 1877, Bastian announced that he was prepared to demonstrate publicly the occurrence of spontaneous generation in neutral or alkaline urine. Pasteur accepted the challenge with apparent alacrity. Bastian then explained, quite reasonably, that he wished to limit the debate to the 'single question' of whether uncontaminated potash in urine could create the necessary conditions for spontaneous generation to occur. Like Pouchet, Bastian realized that just one demonstration would do the trick. In response, the Pasteur camp dragged its heels until the disgruntled Bastian got so fed up that he re-crossed the Channel without having performed a single experiment. Once again, and quite inappropriately given their position within the debate, Pasteur and his allies refused to play the game unless they were allowed to select the field where battle was joined. Revealingly, although Pasteur publicly ascribed Bastian's results to sloppy methodology, in private he and his team took him rather more seriously. As Gerald Geison's study of Pasteur's notebooks has recently revealed, Pasteur's team spent several weeks secretly testing Bastian's findings and refining their own ideas on the distribution of germs in the environment.

As we now know, however far they fell short of what might be expected of a hero of science, Pasteur's tactics saved him from almost certain disaster. Had he publicly replicated the experiments of either Pouchet or Bastian he is likely to have produced some hard-to-refute evidence in support of spontaneous generation. Almost certainly, Bastian's potash solution and Pouchet's hay infusions were contaminated, but the conceptual framework within which they and Pasteur worked meant that the contaminant was most unlikely to be identified. All three shared the belief that no form of organic life could long withstand an environment held at the boiling point of water. This, however, is a false assumption. Some microbes can only be destroyed by heating under pressure to 160 °C and then subjecting them to a cycle of repeated heating and cooling.

The yeast solutions used by Pasteur were unlikely to contain heat-resistant bacteria, but this was not the case with potash or hay. In fact, it is almost certain that Pouchet's hay was infected with the bacterium *Bacillus subtilis*. This amazing bacterium can survive extremely high temperatures and will increase in numbers rapidly on exposure to oxygen. The sterilizing precautions of neither Pasteur nor Pouchet could eradicate these tenacious organisms. So had Pasteur returned to the Mer de Glace and used hay instead of sugared yeast-water, he might have been faced with an awkward dilemma: publish his apparent vindication of Pouchet or quietly suppress the results on the pretext of an inadvertent contamination he could not explain. If he had done the honourable thing by accepting Pouchet's results, and subsequently struck on the idea of unusually heat-resistant bacteria, there is every likelihood that he would have been told to go away and stop making excuses. Either way, until this possibility was understood, the debate over spontaneous generation could not be empirically settled.

Spontaneous generation equals evolutionism equals heresy

To dig yet deeper into the question of why the great mass of French establishment opinion-formers were so ready to accept Pasteur's flawed science, we need to consider the wider societal implications of the Pasteur/Pouchet debate. The key lies in the links spontaneous generation had to other ideas that rendered it—in the Establishment's minds—profoundly disturbing. It is now hard to grasp that a concept that today seems quaintly perverse was once believed capable of threatening the stability of the state. Nonetheless, this is a vital element of the context in which Pasteur and Pouchet joined battle.

Not everyone who attended Pasteur's Sorbonne lecture in 1864 arrived with the innocent expectation of learning the unambiguous truth. During the years of the Enlightenment, the idea of spontaneous genera-tion had become inextricably linked to the notion of evolution. One of the first evolutionists was the naturalist Jean-Baptiste Lamarck. In the last years of the eighteenth century, Lamarck had dared to challenge the words of Genesis, '*And the Lord God formed man of the dust of the ground, and breathed into his nostrils the breath of life; and man became a living soul*'. Instead, he claimed that new life arises continually and spontaneously before embarking on a preordained pathway of progress from simple monads into complex forms such as *Homo sapiens*. Lamarckism thus deprived the Bible's Creator-God of even a caretaking role. But in a society that had been torn apart by decades of revolution in which atheistic ideas had been used to help galvanize the middle and lower classes against the monarchy, aristocracy, and the Church, it is small wonder that Lamarck was ostracized when the forces of reaction regained power. In the end, his career was intentionally destroyed by the dogged and resourceful efforts of Georges Cuvier, an arch-conservative, and the greatest naturalist of the early nineteenth century.

During the 1860s the French Roman Catholic Church once more wielded immense spiritual and political power and was again dedicated to suppressing heresy. Not least because Emperor Louis Napoleon had secured the throne with the help of the Catholic Church, attacks on scriptural accounts of Genesis were guaranteed simultaneously to raise political as well as religious storms. As Richard Owen, Britain's finest contemporary physiologist, pointedly remarked, Pasteur's experiments 'had the advantage of subserving the prepossessions of the "party of order" and the needs of theology'. Well aware of all this, Pouchet made every conceivable effort to deny the atheistic implications of spontaneous generation. Predictably such efforts were to no avail. By 1858, the perceived associations between spontaneous generation, evolutionism, and atheism were too strong to be broken. To the Establishment, spontaneous generation and atheism were synonymous, and both had the unmistakable reek of sedition. In such a context, neither Pouchet nor Bastian was ever going to receive a fair hearing.

It was customary during the mid-nineteenth century for a formal commission appointed by the Académie des Sciences to settle protracted scientific disputes. There can be no doubt that the commission set up in 1863 to adjudicate between Pasteur and Pouchet was deliberately stacked against the unfortunate Rouen naturalist. During the 1860s and 1870s a conservative—if highly accomplished—clique did all it could to uphold the veracity of Pasteur's experimental proofs. As we have seen, it did so against mounting evidence seemingly supportive of the rival view. But the only consequence of a strengthening of Pouchet's position was a commensurate increase in the bias of the commission. Seemingly having evinced spontaneous generation in the Pyrenees, in 1864 Pouchet secured the appointment of a second Académie commission. Unfortunately for him, by this time Pasteur's friends exercised almost complete control over its pronouncements. More than this, Louis Napoleon had recently made one of Pasteur's closest supporters in his contest with Pouchet a senator and minister of agriculture. Apprised of these developments, and convinced that he would receive a partisan hearing, Pouchet withdrew his challenge and quit the field.

Since embarking on his debate with Pasteur, Pouchet had also come to recognize that the commission's attitude was shared by the vast bulk of the French scientific elite. By an accident of timing, this august body was then deeply committed to refuting the potentially atheistic evolutionary ideas advanced across the Channel in Charles Darwin's *On the Origin of Species*. Such were the associations between ideas of spontaneous generation and evolution that Pouchet found himself virtually friendless among an elite without whose support no theory could be judged on its merits. By 1869, even discussion of spontaneous generation was prohibited at the famous Muséum d'Histoire Naturelle in Paris. And when Felix Pouchet's son, George, protested against this decree, he was summarily stripped of his position there as *aide-naturalist*. Paternal loyalty ended his career before it had even begun. Thus, despite outward appearances, the contrasts between French political and scientific debate were not dramatically different: the Académie had become little more than an extension of the Second Empire senate.

Few hero-makers can avoid the temptation of claiming that their heroes had to fight ignorant prejudice before having their ideas accepted. Pasteur's hagiographers are no exception. This fairly standard example is from Frank Ashall's *Remarkable Discoverers* (1995):

> In the face of opposition to his ideas, [Pasteur] eventually persuaded the French Academy of Science to appoint a committee to repeat his experiments so that they could be verified. His confidence in his own data was unfailing, whereas his opponents withdrew their opposition, obviously because of their lack of certainty in their own data. Spontaneous generation was vanquished once and for all.

In the light of what has been said in this essay one can see by how much the record has to be distorted to make Pasteur's case fit Ashall's 'hero of science' mould. But to point out that the Académie des Sciences was a rigged jury is not necessarily to implicate Pasteur in its dealings. Might he not have been an innocent beneficiary of the prejudices of others? Evaluating this possibility requires a deeper exploration of Pasteur's private world view.

The preconceptions of Louis Pasteur

Louis Pasteur's father, Jean-Joseph, had fought with striking bravery in the Peninsular War. Whilst serving in Napoleon's formidable Third Regiment he earned the cross of the Legion of Honour and the rank of sergeant-major. Although after the war he followed in his own father's footsteps as a tanner in Arbois, Jean-Joseph looked back wistfully on the days of Napoleonic glory and tried to imbue his son with the same loyalties to French militarism and strong government. Perhaps Jean-Joseph also looked on his son to fulfil the ambitions that had once seemed attainable by himself in the days when the rest of Europe shrank before France's might. Initially, as is so often the case, his wilful son reacted against his parents' values. During the restoration of the Second Republic in 1848 the young Louis donated all of his savings to the Republican cause and joined the National Guard. As his parents fretted at home in distant Arbois, their son was risking his life for an alien cause.

Yet, in retrospect, Pasteur's early commitment to Republicanism has the air of a youthful amorous adventure that was followed by a comfortable and conventional marriage. By the 1860s he had fully embraced the moral and political code of his petit bourgeois stock. As an eminent Parisian scientist, he was now resolutely conservative and immensely proud of his loyalties to the reactionary Louis Napoleon. Nor was his outlook much altered by Napoleon III's military defeat and abdication. Seeking election to the Senate in his home town in 1875, Pasteur campaigned vigorously on the ticket that he would 'never enter into any combinations whose goal is to upset the order of things'. This manifesto was hardly borne of political pragmatism, for in the subsequent elections he was comprehensively defeated.

Moreover, if Pasteur's politics were solidly bourgeois, his religious views were no less orthodox. A series of laboratory notebooks, only recently made available to the public, make quite clear that Pasteur had a non-negotiable—if unsophisticated—belief in a Creator-God. During the 1860s he undertook a series of investigations into the differences between organic and inorganic matter. In his notes he repeatedly insisted that only the Creator-God had ever exercised the power to convert the inanimate into the living. The possibility that life could be created anew without man first discovering the secrets of the Creator was rejected without any attempt at a scientific justification. Pasteur's political and religious conservatism, combined with his close ties with the emperor, had made it virtually impossible for him to subscribe to the ideas of Felix Pouchet. In the climate of the 1860s, passionate support for Louis Napoleon and the status quo were simply incompatible with the advocacy of spontaneous generation.

This backdrop of preconceptions indicates why Pasteur opened his Sorbonne lecture with a discussion of the religious and political implications of spontaneous generationism:

> Thus, gentlemen, admit the doctrine of spontaneous generation, and the history of creation and the origin of the organic world is no more complicated than this. Take a drop of sea water . . . and in the midst of this inanimate matter, the first beings of creation take birth spontaneously, then little by little are transformed and climb from rung to rung—for example, to insects in 10 000 years and man at the end of 100 000 years.

One can picture him pausing for a second at this point and then, donning the garb of the disinterested man of science, he announced with all due *gravitas*, 'But . . . Neither religion, nor philosophy, nor atheism, nor materialism, nor spiritualism has any place here'. These words now ring very distinctly hollow. With access to his notebooks, it is now virtually impossible to believe that Pasteur left his explicitly conformist political and religious views at the door of his institute every morning. Considering that he suppressed a considerable amount of negative data and refused to replicate key experiments, there is every reason to believe that Pasteur's politico-religious convictions *did* strongly influence his interpretation of his experimental data. A man who labels experiments seeming to evince spontaneous generation as 'unsuccessful', does not approach the question of the origins of life with neutrality.

In sum, the conclusion that the metaphysical overtures and crescendo to Pasteur's Sorbonne lecture were more than elegant flourishes seems unavoidable. In broaching the religious implications of his experiments Pasteur was cutting to the very heart of why his lecture attracted so much attention and why he had gone to such lengths to spoil the retirement of a competent and distinguished provincial biologist.

'The Battle over the Electron'

Millikan's original experiment . . . offered convincing proof that electric charge exists in basic natural units. All subsequent distinct methods of measuring the basic unit of electric charge point to the same fundamental value of electric charge.

'Robert Millikan', *Encyclopaedia Britannica* (1992).

This essay returns to the opening years of the last century and examines work carried out by the most internationally famous American scientist of the 1910s and 1920s, the Illinois-born physicist Robert Millikan. In 1907, having previously enjoyed only modest professional success, Millikan decided to stake his career on a gamble that ultimately led to America's second Nobel Prize for Physics. For the following decade he dedicated himself to resolving one of the most hotly contested issues of early twentieth-century science by finding the answer to another. His primary quest was to discover whether electricity is derived from discrete particles (that is, electrons) or if, as many physicists then believed, it comprises an immaterial pulse of force. Because their minute size meant that Millikan could not hope to see the particles from which he believed atoms to be formed, he had to seek indirect evidence of their existence.

The path Millikan chose was beautifully simple. It involved trying to show that the overall electrical effect is made up of a very large number of tiny points of activity. His breakthrough was to see that if this were the case the differences in electrical charge among different molecules would always be a multiple of the *smallest* value separating two individual molecules. The rationale here is that the tiniest difference in charge found between two molecules must represent the charge of a single electron. So if Millikan's data told him that different molecules had charges of, say,

Left: Robert A. Millikan (1868–1953).

4, 8, 12, 24, 28, and 42, then he could surmise that the unit charge of an electron is 4. Now, he supposed, the successful accomplishment of the second task he set himself—establishing an electron's electrical charge ('e')—would automatically give him the key to his primary quest. If the disparity in charge between two points of electrical activity is always exactly divisible by e, he would have very good evidence that discrete particles (electrons) are indeed the basis of electrical energy

Particles and waves before Millikan

For millennia, hardly a generation has gone by without one or more serious thinkers proposing that the basic building blocks of life comprise infinitesimally small and discrete particles of matter. But only in the past hundred or so years has it been possible to gather empirical evidence to support this intuition. In the final decades of the nineteenth century, it began to seem that a resolution to this fundamental question was at last on the horizon. Several modern-day physicists have reflected that to be a physicist in this period would have been 'very heaven'. The sheer number of new discoveries generated a heady atmosphere of expectancy, and Millikan was quickly caught up in this wave of excitement. He and many others began to realize that electricity, radiation, cathode rays, and X-rays produce physical phenomena that allowed one to determine if they arose out of particulate matter or were some kind of immaterial force. This pursuit acquired immense significance because it seemed reasonable to suppose that if electricity, radiation, and other energy forms are particulate, then so are *all* the elements of nature. Getting to the root of these phenomena, it was agreed, would make or break the atomic theory of matter. In the words of Millikan himself, by 1910 this controversy was the 'most fundamental question of modern physics'.

Much of the data supporting the particulate theory of the electron had emanated from Cambridge University's Cavendish Laboratory and the University of Manchester. In these two institutions, luminaries such as Ernest Rutherford, J. J. Thomson, and C. T. R. Wilson were developing ingenious—if often homespun—apparatus that led the way in the detection of atomic particles. Many of their experiments involved either the study of droplets of cloud water fused around free ions or the firing of cathode ray beams. In 1897, for instance, J. J. Thomson fired energized cathode rays through a glass tube in which the resulting illumination could be observed (as in modern television sets). He found that he could bend and distort the illumination using electrical and magnetic fields. This ability to modulate the behaviour of the rays showed that whatever produced the illumination had an electrical charge, a finding which confounded the idea that such effects were somehow similar to light waves.

None of their findings, however, provided conclusive evidence in favour of the atomic theory. In fact, the well-known Austrian physicist Ernst Mach had good reason sarcastically to demand of a group of physicists discussing the atom, 'Have you ever *seen* one?' In Mach's time, the main explanatory rival of atomism rested on a belief in the existence of 'ether', a medium supposed to transmit electromagnetic waves and fill the entirety of the space around us. Strongly favoured amongst German physicists, this approach drew analogies between the perturbations caused by eddies or ripples passing through water and forces such as electricity passing through the ether. The commonality, as they saw it, was that neither ripple nor electricity had a material form independent of the substance through which it was passing. The origins of both ripple and electrical forces were thought to lay in energetic disturbance of the medium—*not* the discharges of material particles.

Thus, in order to understand Millikan's experiments the historian must exhume a long-forgotten quest that once confounded physicists all over the world: the design of an experiment that would determine the relative strengths of the 'ether' and particulate theories of electricity. First, though, we must be prepared to disregard our present knowledge and treat the ether theory as seriously as Millikan himself did. If we do not, then we will be drawn into thinking that Millikan obtained support for the particulate theory of matter because of its obvious superiority. Because atomic theory triumphed—at least until substantially modified by quantum mechanics—it is easy to overlook just how difficult was the challenge that Millikan set himself.

The importance Millikan attached to showing that the electrical effect was reducible to tiny points of activity arose directly out of the strength of the countervailing view pushed by the ether theorists. As their model offered no explanation for discrete points of electrical discharge all divisible by a unit of fixed value, finding such a pattern would be a major blow in favour of the atomic approach. But Millikan must have realized that he was entering the lists on the basis of winner takes all. Were molecular charges shown to lie on a broad continuum with no regularities, it would provide strong evidence of electricity being no more than a perturbation in the ether and not the result of discharges from individual particles. In short, the whole future of modern physics hung on success, or lack of it, in the detection and measurement of incredibly minute levels of electrical discharge.

The immense difficulties involved in determining the value of e are illustrated by the experiments of Harold A. Wilson. He used the newly devised cloud-chamber apparatus in which air is saturated with water and forms clouds. He also assumed that the water droplets cluster around single charged ions in the glass chamber. On this basis, he proceeded to measure the speed at which such droplets fall: a value determined by the combined effects of gravity, the size of the drop, and the viscosity of the gas—factors calculated with some accuracy by H. A. Wilson. At this early stage of his experiment, the charge of the ion was of no real consequence to the behaviour of the cloud. But Wilson then modified the procedure by subjecting one layer of the cloud to an electrical field that pulled those droplets down to the base of the chamber much more rapidly than those falling under gravity alone. In this way they could make an attempt to work out the charges of the ions themselves. This is because the speed at which the cloud layer gravitated to the anode would be affected by the amount of charge its droplets carried. H. A. Wilson finally came up with a mean figure for e of 3.1×10^{-10}.

There was never any suggestion, however, that this had been a crucial experiment. This is because the researchers were unable to eliminate dozens of complicating factors. The most important of these was the rate at which water in the droplets evaporated. To Wilson and his fellow atomists this made it necessary to assign a very large margin of error to the mean value, ranging from 2.0×10^{-10} to 4.4×10^{-10}. Yet, although H. A. Wilson felt that he was nudging closer to gaining an accurate value for e, he realized that his results were being seized on by ether theorists as being fully supportive of *their* case. Instead of accepting that the spread of results reflected the effects of confounding variables, ether theorists saw them as clear evidence of a continuum of electrical effects consistent with the idea of electricity involving disturbances to the electromagnetic ether. In short, stalemate.

Similar experiments by Thomson, Rutherford, and C. T. R. Wilson suffered from the same defect. The imprecision of their experiments meant that they could offer a value of e derived only from statistical averages that did not even form a neat statistical distribution. To the atomists this seemed reasonable; to ether theorists the average values thus obtained for e were mere artefacts of the statistical methodology. In their view the highly variable results themselves constituted the key evidence, being precisely what ether theory predicted. Only one thing was going to resolve this impasse in favour of the atomists: outstanding experimental results tightly clustered around a consistent value for e that was itself perfectly divisible into all higher values obtained. This was the very tall order that Robert Millikan set out to deliver.

A stroke of luck and a breakthrough

No one would deny that Millikan approached his task with clear pre-conceptions. He was already committed to the particulate view and so straightaway focused his efforts on Rutherford's suggestion that the earlier work had been confounded by the evaporation of water in the cloud chambers. Stopping evaporation was exceptionally difficult, how-ever, so Millikan elected instead to measure its rate with a view to then factoring this into the results he would obtain by rerunning the earlier experiments. He began by attempting to use an electric field to hold the top layer of the cloud steady—against gravity—so that he could deter-mine the rate of evaporation undisturbed. The result, however, was an experiment that seriously but very fruitfully backfired. Sending 10 000 volts through the top of the cloud did not hold it still. Instead, Millikan was shocked to witness a large part of the cloud dissipate almost com-pletely as, destabilized by the field, droplets repelled and raced away from each other.

Millikan's brilliance lay in immediately appreciating the significance of this observation. The earlier experiments of Rutherford, Begeman, and the Wilsons had concentrated on the clouds themselves and not on

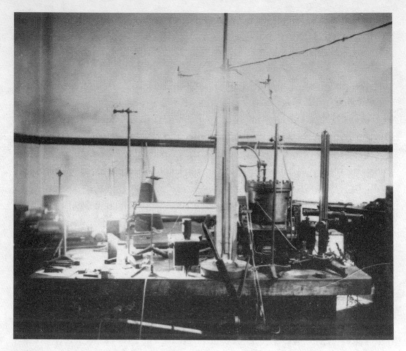

Millikan's oil–drop experimental apparatus.

individual droplets. It had been uncritically assumed that each of the droplets had the same charge. But what Millikan witnessed was a scene analogous to the results of an earthquake in a built-up area. Although most of the buildings immediately disintegrate, a few have the necessary structural qualities to remain standing. Similarly, in Millikan's cloud chamber although most of the droplets had disappeared, a limited number remained.

These few suspended droplets, Millikan reasoned, had exactly the mass and charge necessary for the field to counteract the effects of gravity as originally intended. Rather than being made up of droplets with a uniform charge, then, the original cloud had held particles with a wide range of charges. The cloud's residue, however, had another lesson to impart. Millikan explained in his subsequent scientific article of 1910 how these single droplets, frozen in place, provided the 'first definite, sharp, unambiguous proof that electricity was definitely unitary in structure'. His reasoning for this was simple. A match between the individual charges on a number of droplets and the countervailing forces so perfect that it held them steadily in place spoke eloquently of a standard scale of electrical charges. An infinitely variable range of perturbations within the ether was most unlikely to have achieved such consistency.

Millikan's new evidence was not limited to this observation. By varying his electric field, he now had the means of selecting individual droplets and establishing the necessary voltage to hold them at rest. When released, their rate of unimpeded descent could be used to determine their mass by applying formulae derived from the differential effects of air resistance on spheres of different sizes. Used in conjunction, these two figures enabled him to calculate the overall electric charge on each droplet. Having done this experiment many times, Millikan had the great satisfaction of noting that the relationship between charges was just as predicted by atomic theory. 'Charges actually always came out', he later wrote, 'easily within the limits of error of my stopwatch measurements, 1, 2, 3, 4, or some other exact multiple of the smallest charge on a droplet that I ever obtained.' This smallest figure, then, was the charge of a single—particulate—electron. Even more striking, during his experiments, when a droplet was suspended in the chamber it was often possible to see it shift its place within the electrical field. Millikan quickly realized that here he was observing atmospheric ions landing on the droplet and altering its charge. 'We could actually see the exact instant at which it jumped on or off.' His sense of excitement is almost palpable.

Guilty as charged

In February 1910, Millikan's description of his new method was published in the prestigious *Philosophical Magazine*. Asserting a value of e of 4.65×10^{-10}, he presented the data on which this figure was based and, for the first time, physicists realized their folly in overlooking the possibility of studying individual water droplets. Stylistically, however, this paper was most unusual. Experimental physicists (just like science students at school) naturally make decisions as to which experimental runs are worthwhile including in their final summary of results. Some experiments go simply and obviously wrong and no one demurs at their being overlooked. In other cases the result itself is so aberrant that error seems the most reasonable explanation. Of course, trying to work out why an unusual result has appeared might be the first step on the road to a Nobel Prize. It is much more likely, however, to prove a time-consuming route to the uncovering of the way in which one of the researchers failed to follow the appropriate method. This is how the American geneticist Theodosius Dobzhansky put it:

> Few experimenters are lucky enough to have no mistakes or accidents happen in any of their experiments, and it is only common sense to have such failures. . . . In view of the high frequency of bogus results, a perfectly legitimate convention has arisen among scientists by which most inexplicable results are casually suppressed.

As Dobzhansky's words indicate, the suppression of unexpected results is not necessarily bad science. Most of the time the risks of the baby sharing the fate of the bathwater are low. Nonetheless, frank admissions of selectivity are rare. The vast majority of scientists instinctively prefer to uphold the image of themselves conscientiously following wherever the data lead them. At least with his first paper, Millikan was a striking exception. With little prior experience of publication, he was very much the innocent abroad. His article included numerous scores from his dataset that he considered to have come from unsatisfactory experimental runs. Where his peers would have either jettisoned the results or included them without caveat, Millikan marked each run with one, two, or three stars depending on how well he thought the experiment had gone. His calculation of an average score for *e* then involved differentially weighting his results according to the number of stars awarded.

No doubt with the very best of intentions, Millikan was publicly broadcasting a willingness to 'tamper' with his results after they had been produced. There could be no suggestion of dishonesty as he was being so open about his procedure. Indeed, given the technical difficulties of cloud-chamber experiments, this made considerable sense. His strategy did not, however, conform to the way in which leading-edge science is reported, then or now. One clause in particular left many readers uneasy.

Explaining why he had included several estimates of e derived from apparently unsatisfactory experimental runs, he noted almost as an aside, 'I would have discarded them had they not agreed with the results of the other observations'. He then proceeded to explain on what bases he had entirely ignored seven experimental results. In one case he explained, 'I have discarded one uncertain and unduplicated observation apparently upon a single charged drop, which gave a value of the charge on the drop some 30 per cent. lower than the final value of e'. Although there is a pleasing innocence in this, it does reveal a rather poor grasp of what is usually thought to be the proper relationship between research findings and research conclusions.

Felix Ehrenhaft's critique

Clearly, as the philosopher-historian Gerald Holton has shown, Robert Millikan was not dispassionately accepting what his data told him. And although he may have been vindicated in the following months and years, the experiments that made his name only 'became' conclusive once further evidence had accumulated from several different directions. The subjectivity of Millikan's approach is nicely illustrated by examining the work of one of his fiercest critics: the Austrian physicist Felix Ehrenhaft (1879–1952). Ehrenhaft's detailed refutation of Millikan's data shows how easily his results could be taken to support a contrary view given different theoretical presuppositions. Ehrenhaft began his work on the nature of electricity as a committed believer in the atomic theory. In 1909, he even claimed that his own estimate of e was more accurate than those so far produced by Millikan. By 1910, however, he had performed a sudden and spectacular volte-face.

Observing the movements of tiny particles of metal and cigarette smoke using an ultramicroscope, and influenced by Continental intellectual traditions, Ehrenhaft began to argue that the low scores for e occasionally recorded by Millikan should be accepted as legitimate results. His own experiments showed a range of e from 7.53×10^{-10} to 1.38×10^{-10}, a spread that he believed reflected the nature of reality itself. So in a devastating critique of the way Millikan had employed 'hypotheses and corrections', in 1910 Ehrenhaft announced that the available data justified a belief in either a theory of 'subelectrons' (particles smaller than electrons) or the alternative theory of electricity as perturbations in the background ether. According to Ehrenhaft, however, it was no longer possible to 'hold on to the fundamental hypothesis of the electron theory'. In fact, the subelectrons that Ehrenhaft claimed to have discovered were the result of weaknesses in his method, but this was not known until some years later. In 1910, it was only Millikan's profound commitment to atomic theory that encouraged him to continue his investigations in the face of Ehrenhaft's stinging attacks.

Millikan had laid himself open to these attacks by being so candid. Ehrenhaft's review made clear to him the error of his ways. An article in *Science* magazine of September 1910 showed that he was learning the merits of discretion. In the meantime, Millikan had also improved his method. By using (non-evaporating) oil drops instead of water he was now better able to show that the charges of ions picked up by the drops from the atmosphere were 'exact' multiples of the original charges of the ions around which the drops had formed. This new method, he enthused, is free from 'all questionable theoretical assumptions' and its value could be understood even by the 'man on the street'. In his *Science* account, although still confessing that several results had been rejected, he made no attempt to weight different scores.

Three years later, in 1913, Millikan's *Physical Review* article showed that his scientific superego was fully operational. His published report claimed a maximum margin of error in measuring e of a mere 0.5 per cent. It also stated, 'This is not a selected group of drops but represents all of the drops experimented on during 60 consecutive days'. Yet, this was not the story told by his private laboratory records when, 70 years later, Gerald Holton began to look at them in detail. It would appear that instead of openly admitting to having excluded poor experimental runs, Millikan was now going to the opposite extreme and falsely denying that he ever had them at all.

More of Millikan's manipulations

The actual process of vetting through which Millikan's data passed is clear from the remarks he jotted in his notebooks after many of his experimental runs. In December 1911, he scribbled 'This is almost exactly right & the best one I ever had!!!'; in February 1912, 'Exactly right' and '*Publish* this Beautiful one'; in March of the same year, '*Publish* this surely / Beautiful!!' and 'Error high will not use'; and in April, 'Perfect Publish', 'Won't work', 'Too high by 1.5%', '1% low', and 'Too high *e* by 1.25%'. In one run, shortly before Christmas 1911, he calculated a value of *e* way outside the expected error margin. In his notebook he coolly jotted, '*e* = 4.98 which means that this could not have been an oil drop [but a speck of dust]'. This highlights the fact that Millikan's use of the word 'beauty' was frequently circular. In each of these cases he had performed the drop, made a hasty calculation of *e*, and scribbled down whether or not the result was printable. And Millikan's notebooks show that rather than publishing every result obtained over a 60-day period, only a third actually made it into the article itself: 117 out of 175 never made it off the laboratory bench. This does not sit at all comfortably with the repetition in his 1917 book *The Electron* of his earlier claim that the drops he recorded in his paper 'represent all of those studied for 60 consecutive days, no single drop being omitted'.

Nevertheless, this is not a straightforward case of scientific fraud. Most of the time Millikan could easily explain why an unusual score had been obtained—the effects of convection currents or contamination with dust—and, generally speaking, recalculating Millikan's *e* score to include his discarded results gives a value of *e* very close to that which he ultimately published. The resultant margin of error is increased but not enough to seriously effect his overall findings. Still, in several cases, the fact that a calculated value deviated strongly from the mean score was enough immediately to disqualify it from inclusion in the published article. In three cases in particular, Millikan derived measures that were very different indeed from his mean value.

On 7 March 1912, he calculated a value for *e* of 1.915×10^{-10}, an alarming 60 per cent outside of his normal range and non-divisible by his typical *e* value. It is exasperatingly hard to find an explanation for this score, and Millikan offered none. According to modern physicists, contamination with dust could not have produced such a discrepancy. Nor does there seem much likelihood that his batteries—and their various backups—simultaneously malfunctioned. What happened with this drop remains a mystery but it was not one that Millikan chose to air. He made no mention of it in his published paper.

On 7 March another anomalous event occurred. On this day he came up with the score $e = 2.810 \times 10^{-10}$. Again, it was enormously wide of the mark and gave a fractional value for the charge of the electron. Interestingly, before the calculation of e had been performed, Millikan was sufficiently satisfied with the way that the experiment had gone to scribble 'Publish' in his notebook. But after he had made the necessary calculations he rejected the data on the basis that 'Something [was] wrong w[ith] therm[ometer]'. Modern researchers agree that this explanation is no more than a rationalization after the event. What really happened is far from clear.

Finally, on 20 January 1912, Millikan took a series of readings on an oil drop that gave incomprehensible results and another score for e way beneath his average. Once more, the result was simply jettisoned.

Positives and negatives

Given the present state of knowledge, Millikan's suppression of these results may just look like tidying up. But it needs constantly to be kept in mind that at the time the article was written it was evidence of the variability of e that Felix Ehrenhaft was using to support both the rival ether theory and the notion of 'subelectrons'. The clustering of Millikan's data around a specific score was uncomfortable for Ehrenhaft, but so long as there were occasional discrepancies—scores that were too low or fractional—then he had breathing space. Where Millikan had good grounds for believing that particular experimental runs had been technically flawed, ignoring the resulting data was entirely reasonable. But this does not apply to the last three cases I have described. These drops were rejected without any plausible technical explanation. And there can be no question that had Ehrenhaft gained access to Millikan's notebooks, he could have published the results of these anomalous drops as lending powerful support to his own theory. Certainly Millikan must have been aware that in suppressing these data he was depriving Ehrenhaft of high-calibre ammunition. To most of the historians who have analysed his notebooks, his tactics were unquestionably underhand. Where Nobel Prizes and international celebrity are at stake, however, this sort of data suppression is unlikely to be rare.

Against this backdrop one can see how inappropriate it is for the loser—Felix Ehrenhaft—to be indicted for lacking the courage to back down. The evidence was simply too ambiguous at the time for us to make *being right* synonymous with *being the finer scientist*. As it is, later impressions have been skewed by a knowledge of 'what came next'. Had Nature really been wired in the way Ehrenhaft claimed, his sheer, dogged determination would by now have made him into a legend just as unreasonably as he is now cast as the misguided also-ran. In this respect, it is instructive to note that Ehrenhaft continued to find both empirical data and journals willing to publish his papers on ether theory well into the 1940s. Nor should one interpret this extraordinary unwillingness to accept the consensus view among physicists as betraying a flawed character. The logic of scientific investigation—constantly trying to disprove accepted theories—was always there egging him on. If he suffered moments of self-doubt, or if he got sick of playing the lone outlaw, he might well have reflected on how few scientific ideas have survived intact and unmodified for more than a decade. Furthermore, whilst it would be wrong to credit Ehrenhaft with having anticipated modern discoveries in particle physics, at the very least we should see that a belief in the probable existence of subelectrons was quite rational.

By way of conclusion, I would like briefly to mention three of the more general morals that can be drawn from this story. First, that the innocuous suppression of data is an entirely unexceptionable part of science. The process of sieving results to decide which should and should not make their way into published texts saves a huge amount of time and prevents resources being diverted into what are almost certainly dead-ends. It may be entirely rational, but this is a practice that just isn't spoken of very much by scientists themselves.

Second, we have another demonstration of the difficulties of keeping experimental expectations and analysis truly separate. It is abundantly clear that Millikan and Ehrenhaft were both relying on prior theory to make sense of nature. But, in a quite unremarkable manner, both were also using ambiguous results to support the theories themselves. Ehrenhaft was prepared to publish all of his results because they were consistent with his belief in a continuum of charge. Millikan, in contrast, was so convinced of the veracity of the atomic theory that he was prepared to treat his data differently depending on the value of e they gave him. Millikan was not privately worried that he was riding roughshod over standard definitions of the scientific method because he believed his theory to be such good science that he was entitled to take some liberties with his data. But had he not been adjudged correct in the long run, it's likely that modern commentators would invoke his story as a homiletic warning against reasoning from weakly attested theories.

Finally, as Gerald Holton stresses, the Millikan story implies that an ability to ignore cogent criticism can play a very positive part in the development of better theories: Millikan's non-inductive belief in the atomic theory insulated him from criticisms at a stage at which a less-determined researcher might have thrown in the towel. Instead, fortified with the atomic theory, Millikan and his fellow physicists went on to collect enormously more compelling evidence for the existence of the electron in the years after 1909. It brings small comfort to the also-rans, but it appears that some degree of what at first glance seems to be irrationality can have an important role in achieving scientific progress.

THE ECLIPSE OF ISAAC NEWTON

Arthur Eddington's 'proof' of general relativity

In 1919 [Eddington] led an expedition to Principe Island
(West Africa) that provided the first confirmation of Einstein's
theory that gravity will bend the path of light when it passes
near a massive star. During the total eclipse of the sun, it was
found that the positions of stars seen just beyond the eclipsed
solar disk were, as the general theory of relativity had predicted,
slightly displaced away from the centre of the solar disk.

'Arthur Eddington', *Encyclopaedia Britannica* (1992).

The expeditions despatched to Brazil and the island of Principe
on the occasion of the total eclipse of the Sun on 29th May,
1919 found that the effect which had been predicted by
Einstein did in fact exist. Quantitatively, too, the agreement is
a good one.

W. Pauli, *Theory of Relativity* (1958).

I magine that having used an exceptionally powerful telescope to determine very accurately the distances between the stars of a constellation,
you repeat the process on another night. On the second occasion it
happens that on its way to you, light from one of the stars is passing very
close to an intervening star or black hole. If unaware of the effect large
heavenly bodies have on light, you would find to your surprise that this
particular star has shifted in relation to its companions in the constellation.
If repeated a third time when, as in the first case, light from all the stars in
the constellation passes nowhere near any stars or black holes, the seemingly errant star would be back where it started. Such apparent movements of fixed stars presents a puzzle, but it is not the stars that create it.
The real cause is the capacity of gravitational fields to warp space-time
and thereby alter the direction in which light beams travel. The degree

Left: Sir Arthur Stanley Eddington (1882–1944).

of distortion depends on the mass generating the gravitational field and how close to it the light beam passes. Neither our brains nor our cameras, however, are configured to take account of such gravitational effects. Instead, when a star beam reaches us after passing very close to a large heavenly body, we instinctively locate the light source by assuming that the light has travelled to us in a straight line. Thus we mislocate the source star.

Unnatural though light bent by gravity may seem, it is not a new idea to science. The possibility that a ray of light is made up of a stream of tiny packages has had supporters for millennia. Once Newton's gravitational theories were accepted, it was recognized that they could have important implications for these hypothesized units of light. On the reasonable assumption that each unit has to have some mass—albeit unimaginably small—then they would be as much affected by gravity as would any other object in the Universe. (To argue otherwise would be to deny the central point that Galileo supposedly demonstrated from atop the Leaning Tower of Pisa.) In 1801, the Bavarian scientist Johann von Soldner calculated just how much deflection one would expect to see. Looked at from a Newtonian perspective, what von Soldner said can be thought of in terms of an imaginary tube through which the beam of light passes on its way to us. Viewed from Earth, this tube can be seen to have three co-ordinates by which to locate the position of any given unit of light at any given stage on its journey: two spatial ones (left/right and up/down) and time. Thus, as our unit of light travels down this imaginary tube, the gravitational pull of any nearby stars or planets can be factored in and the whereabouts of the light unit in space-time calculated with great accuracy.

Or so it seemed until the second decade of the twentieth century. Then Albert Einstein published his ideas on relativity and fundamentally challenged the simplicity of this picture. According to Einstein, it is not the light units that are affected by gravity, but the very time/space co-ordinates hitherto used as absolutes to track their path. Our tube can no longer be imagined as having standard units of space and time throughout its length. Rather, it is as though the reference grid on a map ceases to be an external imposition and becomes, instead, part of the landscape. Like the landscape, it becomes itself subject to the great forces of nature. This is because, Einstein argued, large gravitational fields warp the space-time continuum and, consequently, alter the course of light passing through them.

Even knowing that there is hardly a physicist alive who does not believe in general relativity, these are exceptionally difficult ideas to grasp. But in the first decades of the twentieth century the status of general relativity was that of a clever speculation garlanded by just a few ambiguous observations. Therefore the obstacles that Einstein's supporters faced were immense. Nevertheless, even if Einstein's theory was highly speculative, the history of human thought had rarely encountered such a superbly inventive and reason-defying concept. Before long, physicists on both sides of the fence were racking every available neuron trying to devise methods of testing general relativity against Newtonian mechanics. In 1916, the ante was suddenly raised when Einstein used his theory to calculate that the degree of light distortion caused by general relativity would be roughly twice that predicted by Newtonian physics. To his supporters, this calculation raised the exciting possibility of producing an experimental vindication of Einstein's controversial new theory.

Their opportunity lay in the 1919 solar eclipse. And the challenge they faced was that of measuring a very small effect with sufficient precision to distinguish between Einstein's predictions and the Newtonian alternative. Given the available technology, the Sun was the only body likely to create an effect large enough to be measured from Earth with the necessary accuracy. Usually there was an insuperable difficulty with this. When star beams travel close to the Sun they are completely obscured by its overwhelming luminosity. During a solar eclipse, however, this problem disappears. Because the Moon temporarily obscures the Sun, these star beams briefly enable their source stars to be observed. To take scientific advantage of this, in 1918 two separate British scientific expeditions set out for the tropics. Their plans were to make observations of suitable stars during the eclipse of 29 May 1919 and subsequently to repeat the exercise in the night sky. The expeditions were very well publicized and the scientific community awaited their results at a high pitch of excitement. Towards the end of 1919, a packed meeting of the Royal Society in London finally learned that Albert Einstein's predictions had been fully vindicated. His ascent to scientific pre-eminence was assured and physics would never be the same again.

In the years that followed, it quickly became accepted dogma that these two studies of the eclipse had fully supported general relativity theory. Doubts were occasionally raised, but they were quickly silenced. Now more than 80 years on, '1919' is as symbolic a date for physicists as 1859 (the year of the publication of *On the Origin of Species*) is for biologists and 1776 for constitutional historians. Thus, the British physicist Paul Davies wrote in his 1977 book *Space and Time in the Modern Universe*:

> The bending of light rays by a gravitational field was a central prediction of Einstein's theory, and was triumphantly verified observationally by Sir Arthur Eddington (British 1882–1944) during an eclipse of the sun in 1919, when the bending of starlight by the sun was measured and found to agree with the theoretical value calculated by Einstein.

No doubt the enduring inspirational qualities of these expeditions owe much to their seeming to show the scientific method at its best. First, an innovative theory is developed that challenges an existing paradigm. Second, different predictions based on the same event are derived from the competing theories. Third, exact data are collected and one of the theories justly triumphs over the other. Beyond this, Eddington's story is even more attractive because the 'duel in the sun' he managed to set up refined some very complex physics down to a seemingly simple matter of the degree of deflection. Throw in the exotic locations, the struggle to reach them, and, in counterpoint, the extreme savagery of the First World War, and you have the scientific Odyssey par excellence.

Delve a little deeper, however, and one begins to see that the solar-eclipse expeditions of 1918–19 were no more successful than thousands of lesser experiments—past and present—in satisfying these model criteria. The chief reason that these studies retain their popularity is that Einstein's ideas ultimately triumphed. Looking back on the solar-eclipse expeditions our presentist sensibilities incline us to think that the researchers of 1919 *must* have produced accurate and compelling data. But this, as we have seen in the past two chapters, need not be true at all. Indeed, here again it's clear that the scientists involved were very lucky to be accepted by their posterity as having proved their point. For at the time, as the science historians John Earman and Clark Glymour have shown, the evidence they presented was unquestionably inadequate. This leads on to the further question of why the scientific community embraced with such alacrity an experimental 'proof' that was really nothing of the sort.

Meet the teams

Both 1918–19 eclipse expeditions comprised British physicists. The first team, which observed the eclipse from Sobral in Brazil, was led by A. Crommelin and C. Davidson. The other, headed by Arthur Eddington and his assistant E. Cottingham, made its observations from the island of Principe, which lies off the coast of West Africa. Eddington, born in the English Lake District, was already an eminent Cambridge physicist and it was his interpretation of both teams' data-sets that would serve to vindicate Einstein. For this reason it is noteworthy that even before departing for Principe he was well known for his Einsteinian sympathies. As the most important expositor of general relativity within Britain, most of his colleagues knew that he was undertaking the eclipse expedition in the fervent hope of confirming his radical intuition that Einstein was right.

To understand the difficulties the teams faced we need first to consider the sorts of equipment they used for the task in hand. The Sobral team took with it an 'astrographic telescope' and a 4-inch telescope. Eddington's team took just an astrographic instrument. Their plans, however, were identical. Photograph the star beams close to the edge of the eclipse and then photograph the same stars later in the year in other parts of the sky as a baseline. Crommelin would remain in Brazil to do this, whereas Eddington would return to England and make use of facilities at the University of Oxford.

The teams also took with them the same theoretical predictions. Depending on how great were the displacements found, either Einstein or Newton would be vindicated. They were prepared to endorse Newton if the displacement was in the region of 0.8 second of arc, and Einstein if it was close to 1.7 seconds of arc. This difference is so small that it amounts to measuring less than the width of a penny as seen from over a mile away! This was a tall order indeed. In the event, because there were no stars aligned tightly to the edge of the Sun during the eclipse, they had to settle for ones appreciably further out. As this meant a much weaker gravitational effect, measurement would be proportionately harder. So it is easy to understand why, when the exceptionally accomplished Eddington calculated an arc of displacement close to that predicted by Einstein, he described it as the most 'exciting event I recall in my . . . connection with astronomy'.

The problems

Quite apart from the smallness of the measurements to be made, the technical difficulties facing the two teams were simply immense. The most fundamental problems stemmed from the fact that a comparison was being made between the apparent locations of stars photographed in different parts of the sky in different seasons. Unavoidably, therefore, ambient temperatures were going to differ from one occasion to the other. This is important because the disparity in focal length between a warm and a cold telescope can easily produce a distortion equivalent to that which the experimenters were expecting to observe. A similar effect may be produced by the fact that the solar-eclipse photographs were to be taken during the day and the remaining photographs during the night. Aside from ambient temperature, both studies were also hampered by different degrees of 'atmospheric turbulence'. (This is the distortion to background images, mainly caused by convection currents, that can be seen when looking across the top of a hot barbecue; in tropical locations atmospheric turbulence would have been a very serious problem.) On top of this, both parties faced the unavoidable problem of inclement weather. In the event, clouds were partially to obscure exposures taken by both groups.

Add to these hazards the possible mechanical changes to the telescopes caused by their having to be transported to sites so far from England, when even the slightest damage affecting the angle of the photographic plates would have had disastrous results. Exacerbating this problem, the eclipses had to be observed in remote areas where large state-of-the-art equipment could not be transported. Both teams had to rely on smaller models that required a long exposure time. As such, their telescopes had constantly to be counter-rotated so that the Earth's rotation did not alter the point in the sky at which they were aimed. The mechanisms for rotation that the two teams constructed introduced yet another potential source of error.

Some of these difficulties could be controlled for and taken into account at the calculation stage. This generally involved determining the displacement of stars whose altered position could only have been caused by mechanical changes with the telescopes and photographic equipment. The measure of their displacement could serve as a reliable index to the amount of experimental distortion involved. Once these effects had been quantified, the behaviour of the target star beams could be isolated. But making these adjustments accurately required a minimum of six undisplaced stars in each photographic frame; otherwise there was insufficient data for the statistical procedures to be performed. Additionally, neither team could deny that their experimental method was likely to involve errors that had not been identified and would therefore pass unrecognized.

To give a sense of just how serious these difficulties were, it's worth mentioning that in 1962 a much-better equipped British party tried to reproduce Eddington's findings. At the end of a frustrating attempt to do so they concluded that the method was much too difficult and could not be implemented successfully. In view of the obstacles considered above, this seems far from surprising. The Indian Nobel laureate, Subrahmanyan Chandrasekhar, with whom Eddington had a long and highly personalized academic dispute, later claimed that science was only one reason for the 1918–19 expeditions. He suggested that Eddington's overall leadership was used to obviate his need either to enlist or declare himself a conscientious objector during the First World War. The implication appears to be that this consideration was allowed to out-weigh the known impracticality of the expeditions' objectives. To date, however, there is no independent verification of Chandrasekhar's claim.

The results stage

On the long-awaited night of the eclipse, the Sobral team managed to obtain 19 plates from their astrographic telescope and 8 plates from their 4-inch telescope. Eddington's Principe team was hampered by cloud cover and took away just 16 plates, but only two of these, each showing only five stars, were actually usable. The Sobral team managed to take the clearest photographs with its 4-inch telescope. These suggested a deflection of star beams grazing the Sun at between 1.86 and 2.1 seconds of arc, averaging out at 1.98 seconds. (Note that Einstein's prediction was 1.7 seconds.) The Sobral team's astrograph shots were of a lower quality, but 18 of them were used to calculate an average of 0.86 seconds. In other words, one set of photographs was close to Einstein's prediction, the other was very close to the Newtonian value of 0.8. Unfortunately the first score was too high to be strictly compatible with general relativity and the score in the second set was based on low-quality exposures. In addition, each set of photographs involved very large standard errors. This should have immediately prompted doubts as to the reliability of the averages themselves.

From the usable Principe plates, Eddington calculated a star-beam displacement of between 1.31 and 1.91 seconds. But even these plates were of embarrassingly poor quality and it has been suggested that the mathematical formula he used to reach these figures was in itself biased. Be this as it may, Eddington's two poor plates gave a mean score of 1.62 seconds, marginally below the Einsteinian prediction.

Self-evidently, with such poor and contradictory evidence, attempting a resolution of the controversy on the basis of these figures was an extremely risky affair. Take just one of the hazards mentioned above: atmospheric turbulence. In the hot environments in which both teams were working it was likely that all but the largest displacements would be cancelled out by this phenomenon. Had the teams been measuring star beams just clipping the Sun's edge, their displacement might have been large enough to eliminate atmospheric turbulence as the sole cause. In 1919, however, with the star beams closest to the Sun obliterated by the corona, those that could be observed were some way from the Sun's rim. Consequently the displacements were so small that the entire effect could quite easily have been caused by atmospheric turbulence alone. At some level, the teams were aware of this. Thus, in discussions after the announcement of the eclipse results, Eddington and his assistants admitted that calculations of small displacements were unreliable. Yet, they refused to let this effect their presentation of the measurements. As we have seen, within a few months Einstein's ideas were being adjudged victorious from the pulpit of the Astronomer Royal.

The interpretation stage

The Sobral and Principe expeditions most certainly did not produce measurements that could definitively confirm either Newtonian or Einsteinian theory. In his book *The Physical Foundations of General Relativity* (1972), the British astronomer Dennis Sciama explained that eclipse observations are notoriously 'hard to assess . . . since other astronomers have derived different results from a re-discussion of the same material'. In this case, there can be no doubt at all that both theories could potentially have been declared victorious, although it may have appeared to the Sobral team that the most likely verdict was a tie. But this is not what happened. Under Eddington's hand, the eclipse results were subjected to extensive cosmetic surgery until they matched Einstein's prediction. Without this treatment Einstein could not have been vindicated in 1919.

Eddington began by casting doubt on the scores obtained by the Sobral team. He claimed that their astrographic results were not randomly distributed around the mean score as one would expect with normal data points. Instead, they were mostly beneath it, suggesting that a 'systematic error' had occurred that had artificially lowered the mean score itself. Without this error, he implied, their results would also have approximated to the higher Einsteinian prediction. This was a reasonable argument. The problem was Eddington's abject inability to show that the same error had not occurred in the other data-sets. When challenged, he produced not a single piece of unambiguous evidence to demonstrate that the measurements he accepted were unaffected by the same error. Even more seriously, Eddington conveniently ignored the fact that the Sobral team's astrographic photographs were visually far superior to his own two hazy plates. There may have been valid concerns about the reliability of Crommelin and Davidson's photographs. But one thing should have been clear: Eddington's were very much worse. As the American commentator W. Campbell wrote in 1923:

> Professor Eddington was inclined to assign considerable weight to the African determination, but, as the few images on his small number of astrographic plates were not so good as those on the astrographic plates secured in Brazil, and the results from the latter were given almost negligible weight, the logic of the situation does not seem entirely clear.

This was an understatement of which any Briton would have been proud. Note also that Eddington's two plates contained an insufficient number of undisplaced stars from which to make the necessary adjustments for error (five rather than six). Then factor in the large standard deviation in the results that he accepted (rendering most of the results either too high or too low), and one can understand why Earman and Glymour concluded in their 1980 article, 'the eclipse expeditions confirmed the theory [of Einstein] only if part of the observations were thrown out and the discrepancies in the remainder ignored'. In short, they didn't.

A core principle of the standard model of the scientific method is that theoretical predictions should not be allowed to influence which results are used and which are discarded. In Eddington's approach, however, as with Louis Pasteur and Robert Millikan, predictions and data interpretation became mutually confirming. Eddington evaluated his results according to how they conformed to his preferred theoretical predictions. On one hand, inordinate value was attached to photographs that approximated Einstein's 1.7 seconds of arc deflection; on the other, dubious ad hoc reasons were invented for jettisoning any that disagreed. 'Einstein's prediction had not been verified as decisively as was once believed', Sciama gently pointed out in 1972. Reflecting on eclipse expeditions in general, he added, 'one might suspect that if the observers did not know what value they were "supposed" to obtain, their published results might vary over a greater range than they actually do'. Or, as the Polish-American physicist Ludwik Silberstein said at a meeting of the Royal Astronomical Society in 1919, 'If we had not the prejudice of Einstein's theory we should not say that the figures strongly indicated a radial law of displacement'. So serious were Eddington's manipulations that one strongly suspects that had the predictions of the rival theories been the reverse—Newton high, Einstein low—Eddington would have discarded his own photographs as too hazy and accepted with alacrity the Sobral party's astrographic pictures.

Most of Eddington's contemporaries were either less incisive or less cynical than Silberstein and Sciama. As a result, after careful massaging, Eddington's judiciously selected data-set could be presented as unequivocally supporting his candidate's theoretical predictions. Having discarded a full 18 plates on very specious grounds, he set about writing the official accounts of the expeditions. In these he routinely referred to only two sets of prints: the four 4-inch telescope photographs obtained by the Sobral team and his own very poor two photographs. As these images gave mean scores of 1.98 and 1.671 respectively, few scientific readers could avoid concluding that Newton had been decisively beaten: the reigning champion for over 200 years had fallen at last.

Once the eighteen astrographic plates had been rejected and forgotten, concerns about the quality of the Principe photographs quickly evaporated. The complexities of the issue receded from view, and the controversy between Einstein and the Newtonians suddenly—but falsely—appeared to be a one-horse race. This is clear from the account of the eclipse expeditions in James A. Coleman's best-selling *Relativity for the Layman* (1969):

> The Sobral group found that their stars had moved an average of
> 1.98 seconds of arc, and the Principe group's had moved 1.6 seconds
> of arc. This nearness to the 1.74 seconds of arc predicted by Einstein
> was sufficient to verify the effect.

But in many cases unwittingly, Coleman and dozens of other scientific commentators skate over the fact that among astronomers Eddington's account did not win immediate assent. Already, in 1918, an American expedition had travelled to Washington state to observe an eclipse. They had reported that the 1.7-second light deflection was 'non-existent'. Ten further eclipse observations were made between 1922 and 1952. Only one of these produced seemingly high-quality data, and that suggested a displacement arc of 2.24 seconds—substantially higher than predicted by Einstein. In fact, virtually every eclipse observation was either unreliable or, in most cases, both unreliable and higher than the Eddington scores. In light of these results, many of those at the cutting-edge of research into general relativity sensibly deferred judgement for rather longer than the accepted view implies. Some embraced general relativity only when evidence of an entirely different type became available.

Status and trust

In overwhelming his critics, Eddington used the Royal Society of London to great effect. This body was set up in the late seventeenth century amidst a nation recoiling from a regicide and years of civil war. Against such a background, the peaceful and mannerly resolution of controversies was given a very high priority. Scientists were no exception. The focal point of the Royal Society was a large lecture theatre in which the cream of the scientific establishment could gather to watch experiments being performed. The idea was that members would reserve judgement on any given topic until the relevant experiments had been carried out in front of them. Then, having personally seen the unvarnished facts, the scientific community could democratically arrive at a consensus and thereby avoid protracted conflict. In many ways, the same principles are alive and well today. Journal articles require the inclusion of detailed methodologies that should allow experiments to be repeated in other laboratories. Witnessing and consensus-forming no longer take place within one location on one occasion, but they can nevertheless be achieved.

But there have always been problems with achieving agreement on what an experiment does or does not prove. Today, science takes place on such a vast scale that it is not always convenient to replicate every important experiment performed. Further, as the British sociologist Harry Collins and his American collaborator Trevor Pinch have shown, some experiments require specialized training, highly recondite knowledge, and technical expertise that may take months or years for another laboratory to acquire. This means that scientists sometimes just have to take their colleagues' word for it. In the case of the Sobral and Principe expeditions, quite apart from the tremendous difficulty in understanding general relativity and performing the appropriate calculations, the experiments themselves were exceptionally difficult to perform, extremely expensive, and totally reliant on eclipses of the Sun. Thus, few astronomers were inclined to try to replicate Eddington's results. In these circumstances, most astronomers were more than happy to accept his interpretations without demur. Whatever else it may be, this case is a powerful demonstration of the role of trust in the advancement of science.

Yet, however high Eddington's personal reputation stood in 1919, there were still major challenges facing him. Success required that the scientific community sin by omission by colluding, first, with his suppression of well over two-thirds of the photographs from the Sobral and

Principe expeditions, and, second, with his ignoring the much more equivocal evidence advanced by other eclipse expeditions.

In understanding why the scientific rank and file placed so much confidence in Eddington it has first to be appreciated that, to many, Einstein was already the greatest modern physicist. In addition, Eddington was not only an extremely accomplished astronomer in his own right, but he was British at a time when this counted for a great deal. Taking full advantage of his esteemed status, Eddington had the clout to secure the ascendancy of his own interpretation by enshrining it within a series of seminal papers and books that he himself authored. By 1919, Eddington had also acquired enormous credibility because he was such a fine expositor of general relativity. He grasped its implications with a flair that could not but inspire confidence. Such was his standing in this new scientific area that the following apocryphal story had wide currency. Eddington's fellow physicist Ludwig Silberstein remarks, 'Professor Eddington, you must be one of three persons in the world who understands general relativity'. After a longish pause, he continues, 'Don't be modest Eddington', to which the latter replies, 'On the contrary, I am trying to think who the third person is!' The story is entirely mythical, but it is as illuminating as it is amusing.

Arthur Eddington's apparent vindication of Einstein's ideas also gained rapid credence because of the status of many of its earliest converts. On 6 November 1919, Sir Joseph J. Thomson, the President of the Royal Society, announced to the assembled ranks of the scientific elite, 'It is difficult for the audience to weigh fully the meaning of the figures that have been put before us, but the Astronomer Royal and Professor Eddington have studied the material carefully, and they regard the evidence as decisively in favour of the larger value for the displacement.' With the weight of the President of the Royal Society and the Astronomer Royal on his side, Eddington could hardly have been surprised to read the following banner headlines in *The Times* the following morning:

Revolution in Science
New Theory of the Universe
Newtonian Ideas Overthrown

'It was generally accepted', *The Times* report went on, 'that the observations [of the eclipse] were decisive in the verifying of the prediction of the famous physicist Einstein.' Over the next few weeks *The Times* carried several letters from respected scientists in support of relativity and even one from Einstein himself on the 28 November. The contributions of detractors, in contrast, were invariably scorned. Indeed, if we return once more to J. J. Thomson's announcement, we see that he was as determined to browbeat the scientific community as was *The Times* the general reader. His concluding remarks included the observation that, 'It is difficult for the audience to weigh fully the meaning of the figures that have been put before us'. It seems not unreasonable to paraphrase this as 'It's beyond your competence to judge in this matter so take our word for it'. Thus, if anybody present had challenged Eddington's conclusions, the challenger would have been up against more than the weight of the evidence. With the three-line whip imposed by a seemingly holy alliance of the Astronomer Royal, the President of the Royal Society, and Eddington himself, none saw serious merit in disagreeing.

Once Thomson's decree had been issued, the scientific community accepted the party line virtually en masse. And for the most part they did so despite lacking a proper understanding of the expeditionary data. Clearly, then, in this case much of the scientific community was prepared to endorse interpretations without being able to justify their decision on empirical grounds. Furthermore, most scientists subsequently stood by this position irrespective of the later publication of eclipse data that did not corroborate Eddington's figures. It is extraordinary how little these later critics managed to influence the debate after 1919. Cutting-edge researchers were the only scientists prepared to dispute the Eddington figures, but even though their results were published they did not have the strength to overturn the interpretations of 6 November 1919. After that date, they were battling against what can fairly be called a cultural consensus. Quite rationally, where non-astronomers reached the limits of their knowledge of astronomical science, they followed their instincts and backed their most accomplished and highly regarded colleagues. At least in the short run, what is perceived to be scientific truth is usually to be found on the side of the big battalions.

In matters of gravity, weight counts

The standard story of the eclipse expeditions carries all the hallmarks of presentist history of science. There is a crucial experiment that vindicates a novel and brilliant theory; one man whose foresight and determination permits it to become established fact; and it has the added spice of the experiments being formed in the exotic jungles of Principe and Brazil. (Only a cabal of jealous rivals and an obdurate Church are wanting to make it a classic.) Looked at in the light of modern knowledge, it is so hard to suspend awareness of 'what happened next' that we tend to assume that the results presented in November 1919 amounted to the very best of cutting-edge science. Yet what has now been revealed by historians shows just how lucky Eddington was. Had he not been later vindicated on the basis of much better results, his posthumous reputation would have been severely tarnished and the eclipse expeditions would long since have ceased to inspire undergraduate physicists.

This analysis of the 1919 experiments shows that Eddington fell far short of the canonical rules of the scientific method. More interestingly, it also reveals that there is an inexact correspondence between how closely these procedures are followed and the persuasiveness of the theories that emerge. In 1919, general relativity won the debate because it had the best public relations available. But this was not a new phenomenon. Indeed, there is a certain poetic justice in Sir Isaac Newton having been eclipsed in this way. After all, several recent biographies have shown that it was partly Newton's power-play tactics as President of the Royal Society that managed to win unusually rapid assent for his own ideas two centuries earlier.

VERY UNSCIENTIFIC MANAGEMENT

On Taylor's 'scientific management' rests, above all, the tremendous surge of affluence in the last seventy-five years which has lifted the working masses in the developed countries well above any level recorded, even for the well-to-do. Taylor, though the Isaac Newton (or perhaps the Archimedes) of the science of work, laid only first foundations, however. Not much has been added to them since—even though he has been dead all of sixty years.

Peter Drucker, *Management: Tasks, Responsibilities, Practices* (1973).

According to Peter Drucker, one of the foremost writers on management in the second half of the twentieth century, the American engineer Frederick Winslow Taylor should rank alongside Darwin and Freud as 'a maker of the modern world'. Seen by Drucker as much more important than Karl Marx, Taylor's contribution was to revolutionize industrial management by bringing the cool analytical approach of science to the business of making optimal use of staff, machines, and materials. There is a sense in which this assessment of Taylor's importance is shared by his sternest critics. For those who give emphasis to the human side of management, Taylor offers such a perfect antithesis as to make him someone who, had he not existed 'would have to have been invented'. Even in his own day Taylor was a figure of fear and abomination. So great was the antagonism of the trade unions towards his approach to industrial management that they successfully pressed the House of Representatives to set up a special committee to look into it. The depth of their concern is well caught in this question put to Taylor by the chairman of that committee, in 1912:

Left: Frederick Winslow Taylor (1856–1915), 'the father of scientific management'.

> Under scientific management, then, you propose that because a man
> is not in the first class as a workman, there is no place in the world for
> him—if he is not in the first class in some particular line, that he must
> be destroyed and removed.

In fact, there is no strong evidence that Taylor ever thought in such apocalyptic terms. His ideas operated at the level of the individual enterprise, with the wider economy tacitly assumed to be somehow dealing with those who found no place in his scheme of things. When he did address the question of the fate of those who did not fit in, he displayed a mixture of self-delusion and pity. Speaking of work carried out in a bicycle ball-bearing factory by a colleague, he had this to say about the implementation phase:

> For the ultimate good of the girls as well as the company . . . it became
> necessary to exclude all girls who lacked a low 'personal coefficient'.
> And unfortunately, this involved laying off many of the most intelli-
> gent, hardest working, and most trustworthy girls merely because
> they did not possess the quality of quick perception followed by
> quick action.

But even such a brief flicker of humanity is a rarity in Taylor's writings. For the most part he works as assiduously as his critics to build up an image not unlike that of Star Trek's 'Spock', save only for the emotional 'lapses' attributed to genes furnished by Spock's Earthling mother. Seemingly a ruthless calculating machine, Taylor's approach to employees was just like his attitude to machines. He had, for example, optimized the durability of industrial drive-belts by determining experimentally the ideal tension for any given size. Then, by presenting his findings in tabular form, he enabled workers, much less skilled than previously, to get it right every time. Introducing cold logic into production and the management of labour was Taylor's *raison d'être* throughout a long and spectacularly divisive career.

The awe in which F. W. Taylor is still held today owes a lot to the astonishing results he cited in his magnum opus, *The Principles of Scientific Management*. This best-selling book, still in print almost a century after it was first published, comprises a stream of reasoned invective against the primacy of 'habit', 'rules of thumb', and 'hunches' in industrial practice. Probably the best-known example in this book concerns the burly gangs of men who were employed by the Bethlehem Iron Co. of Pennsylvania to load bars of pig iron into railway trucks. Throughout the twentieth century, textbooks on organizational theory, industrial psychology, and kindred disciplines almost invariably included some reference to how Taylor raised the productivity of these pig-iron loaders by more than 300 per cent in just a few weeks. The enduring popularity of his account lies in the way it so clearly illustrates the three principles Taylor said underpinned his 'scientific' approach to management. First, the 'careful selection of the workman', or finding out who was fit to join the 'aristocracy of the capable'. Second, inducing the workman to accept a carefully set piece-rate employment. Third, undertaking a detailed study of the ways in which a workman's approach to production could be rendered more efficient.

His ideas long out-lived him. During the 1920s and 1930s Taylor emerged as an icon of modern, cost-effective management. His fan club was not restricted to business leaders and managers in capitalistic democracies: both Lenin and Mussolini are said to have found his ideas to be of great value. Further, his belief in the necessity for educated professionals to study industrial production and promote rational improvement has helped spawn an entire industry of management consultants. These new 'Masters of the Universe' follow in Taylor's footsteps and, fittingly, the principles they learn either at business school or during in-house training are often explicitly tied back to Taylor's original studies. Yet, the man that Drucker has described as responsible for 'the most lasting contribution America has made to western thought since the Federalist papers', is a rather more complex character than the above account suggests.

Quite regardless of the general inappropriateness or otherwise of his methods in the workplace, Taylor's seemingly groundbreaking research in the Bethlehem iron works fell short of even the most generous and inclusive definition of the scientific method. In no small part this is due to the account he gives of what actually happened being false in almost every detail. We know this because more than 60 years after the Bethlehem research (during which period Taylor's story became widely accepted), one of the two men who had actually undertaken the studies—Hartley C. Wolle—died on his farm in Princess Ann, Maryland. Among his effects was found a copy of the report that he and James Gillespie had presented

to Taylor on 17 June 1899, and on which the latter based his most important conclusions. The profound discrepancies between this document
and the account contained in *The Principles of Scientific Management*
fundamentally alters our perception of a man always described by his
biographers as blessed with conspicuous 'integrity'.

'A little Pennsylvania Dutchman'

At the Bethlehem Iron Co., pig-iron loading was exhausting, back-
breaking work. Gangs of men carried heavy pigs of iron up planks and
stacked them in waiting railway cars ready for shipment. Partly because
the work was tough, monotonous, and only modestly remunerative, F.
W. Taylor and his colleagues expected there to be a vast disparity between
potential and actual output. In 1898, two aspiring scientific managers,
Gillespie and Wolle, were assigned to observe the loaders. Their full task
was to provide an 'object lesson' for the benefit of all Bethlehem workers,
in how much more men could earn if they worked under a piece-rate
system. Their first job, therefore, was to set a piece-rate that was both fair
to the workers and likely to promote harder work. Gillespie and Wolle
established that, paid a daily rate, each man loaded an average of 12.5 tons
of pig iron per day. They then selected 10 men from a gang of well-built
Hungarians under the foremanship of a certain John Haack. These tough
men were instructed to load pig iron 'at their maximum speed'.

Fourteen minutes after starting they had loaded 16 tons—more than
they usually loaded in an entire day! Hardly surprisingly, though, they
were soon 'utterly exhausted'. From this and other studies, Gillespie and
Wolle decided that a really 'first-class man could load 7.5 tons per hour at
maximum rate. Using this figure they established the weight that the pig-
iron loaders 'should' be loading every day. Having reduced the hourly
rate by 40 per cent to allow for rests and unavoidable delays, they calculated that each first-class man should be loading 45 tons per day. This
figure was then used to calculate the 'appropriate' piece-rate, which
worked out to be $0.0375 per ton. So to take home the wage they had
received under the daily-rate system, the loaders would now have considerably to increase their productivity. Only those exceeding the new
45-ton minimum would take home more.

In *The Principles of Scientific Management* Taylor claims that having established the $0.0375 hourly piece-rate, his next step was 'the scientific selection of the workmen'. Not all men, he explained, are equally inclined to take advantage of piece-rates and a good pig-iron loader must have the proper mental attitude. Matching the man to the job was a central part of his philosophy. In the case of the pig-iron handlers, he says:

> A careful study was then made of each of these men. We looked up their history as far back as practicable and thorough inquiries were made as to the character, habits, and the ambition of each of them.

Taylor illustrated what he was looking for in this case by immortalizing a 'little Pennsylvania Dutchman' (a term then applied in the United States as much to immigrants of German stock as to those who came from The Netherlands) whom he renamed 'Schmidt':

> I took a labourer who was handling pig-iron and tried to get him to handle more. I took one single man. He was a very quick and wiry fellow, but there was another reason. I knew he was building a house and that he needed money, and that he was independent, and that he was anxious to succeed. I persuaded that man that he could handle instead of fifteen tons a day, forty-five tons.

On wages of only $1.15 a day, Taylor wrote elsewhere, Schmidt was 'engaged in putting up the walls of a little house . . . in the morning before starting to work and at night after leaving'. Not only that, he was known for 'trotting' home for over a mile in the evening 'as fresh as he was when he came trotting down to work in the morning'. The description of Schmidt displays all the Victorian, Puritanical values that the Quaker-born Taylor expected in a good, loyal workman. Furthermore, Schmidt was enthusiastic about making money—'a penny looks about the size of a cartwheel to him', Taylor reports one of his colleagues to have said of him. But not all of Taylor's description was so harmlessly quaint. Taylor also says that Schmidt conformed to his belief that 'a man who is fit to handle pig iron as a regular occupation [must] be so stupid and so phleg-matic that he more nearly resembles in his mental make-up the ox than any other type'.

Satisfying all these credentials, Schmidt became the exemplar of Taylor's ideal pig-iron loader. In *The Principles of Scientific Management*,

Taylor detailed how he had managed to select several more of what he called 'high-priced men'. He explained that he had word spread beyond the confines of the Bethlehem plant that he was looking for men prepared to earn a high wage for which they would have to work hard. Men who came forward could then be scrutinized and their mental and physical fitness for the work scientifically evaluated. Taylor later described how the local press got wind of his plans and went on the offensive. But 'the newspapers,' he noted rather smugly 'even in ridiculing us, did us the greatest service . . . and gave us the best advertisement all over'. Apparently, the only effect of their opposition was to attract considerable attention to Taylor's experiment and thereby enable him to recruit all the 'high-priced men' he required.

The piece-rate

The next stage in the procedure, described by Taylor, was that of persuading the men he had selected, 'to work according to the scientific method'. This meant first getting them to accept a piece-rate system: no small feat where attempts to introduce piece-rates generally resulted in strikes and mass walk-outs. The workers feared that they would cease earning whenever they became ill or very tired. Taylor, however, seems to have had little difficulty with this. His approach was to emphasize the benefits to the worker; in the case of Schmidt he seems to have been able to recall the entire conversation. Couched in arrogant, patronizing terms, the exchange he reports seems almost designed to fulfil the worst expectations of Taylor's severest critics:

> 'Schmidt, are you a high-priced man?'
> 'Vell, I don't know vat you mean.'
> 'Oh yes, you do. What I want to know is whether you are a
>   high-priced man or not.'
> 'Vell, I don't know vat you mean.'
> 'Oh, come now, you answer my questions. What I want to find out is
>   whether you are a high-priced man or one of these cheap fellows
>   here. What I want to find out is whether you want to earn $1.85 a
>   day or whether you are satisfied with $1.15, just the same as all those
>   cheap fellows are getting.'

'Did I vant $1.85 a day? Vas dot a high-priced man? Vell, yes, I vas a high-priced man.'

'Oh, you're aggravating me. Of course you want $1.85 a day. Every one wants it! You know perfectly well that that has very little to do with your being a high-priced man. For goodness' sake answer my questions, and don't waste any more of my time. Now come over here. You see that pile of pig iron?'

'Yes.'

'You see that car?'

'Yes.'

'Well, if you are a high-priced man, you will load that pig iron on that car to-morrow for $1.85. Now do wake up and answer my question. Tell me whether you are a high-priced man or not.'

'Vell—did I got $1.85 for loading dot pig iron on dot car tomorrow?'

'Yes, of course you do, and you get $1.85 for loading a pile like that every day right through the year. That is what a high-priced man does, and you know it just as well as I do.'

'Vell, dot's all right. I could load dot pig iron on the car tomorrow for $1.85, and I get it every day, don't I?'

'Certainly you do—certainly you do.'

'Vell, den, I vas a high-priced man.'

'This seems to be rather rough talk', Taylor went on. 'But for men of the lower mental type like Schmidt, speaking down to them is both necessary and effective in fixing [their] attention on the high wages which [they] want and away from what, if it were called to [their] attention, [they] probably would consider impossibly hard work.' The result? Under Taylor's 'scientific' guidance, Schmidt began to load 47 tons a day instead of the old rate of 12.5 tons. He 'practically never failed' to maintain or exceed this level 'during the three years that the writer was at Bethlehem':

> Throughout this time he averaged a little more than $1.85 per day, whereas before he had never received over $1.15 per day, which was the ruling rate of wages at that time in Bethlehem. That is, he received 60 per cent higher wages than were paid to other men who were not working on task work.

Having shown the effectiveness of scientific selection and carefully packaged inducements on Schmidt, the same tactics were used on the other pig-iron loaders. The results, Taylor boasted, were equally gratifying for the scientific manager:

> One man after another was picked out and trained to handle pig iron at the rate of 47½ tons per day until all of the pig iron was handled at this rate, and the men were receiving 60 per cent more wages than other workmen around them.

Thus, simply by replacing a pig-iron loading gang selected for physical capacity alone with a new, trained team recruited on the basis of both strength and a conducive mental approach, Taylor increased the rate of pig-iron loading from 12.5 tons to in excess of 45 tons per day. As this episode seems to be nothing less than a triumph on Taylor's part, it is hardly surprising that it has found such a revered place in the management literature. Yet as the North American business scholars Charles D. Wrege and Amedo G. Perroni discovered on analysing the accounts of Gillespie and Wolle, Taylor's reconstruction has only the loosest connection with the facts.

Gillespie, Wolle, and the piece-rate

F. W. Taylor clearly implied that he had been involved in the day-to-day running of the studies. Far from crediting Gillespie and Wolle, he wrote them out of history. On the other hand, this was by no means straightforward plagiarism as Gillespie and Wolle's report differs in all important respects from Taylor's version of events. The primary difference is that the pig-iron studies at Bethlehem Iron Co. never involved attempts at scientific selection. The above account is almost entirely a fiction dreamed up by Taylor himself. In reality, strong resistance by most of the workers meant that, in attempting to introduce piecework, Gillespie and Wolle had to accept virtually any individual prepared to accept their terms. Only two aspects of their report directly support Taylor's later claims. First, an attempt to introduce piece-rates was a central issue; and, second, the efforts of one particular worker were crucial to their ultimate success. Regarding the latter, a search for 'Schmidt' during the 1970s proved that this was the pseudonym for one Henry Noll, who now lies buried in the fireman's plot at the Bethlehem Memorial Park. Almost everything said in relation to him and his colleagues by Taylor is, however, false.

In their official report, Gillespie and Wolle described how they started to implement piecework among the pig-iron loaders on 16 March 1899. No mention was made of scientific selection at this juncture or at any other stage in their report. Their job was simply the effective introduction of performance-related pay. They began with 10 men drawn from the Hungarian gang of John Haack. These men were uncompromisingly told that the following day they would start working at the rate of $0.0375 per ton. This meant that their rate of work would have to increase significantly if they were to take home a reasonable wage. Symptomatic of the naiveté of these young researchers, when they turned up for work at 7.30 a.m. the following morning they were surprised to find that the men were loading pig iron on day work and not piecework. John Haack, the foreman, explained that the loaders had collectively refused to work the piece-rate and that he had had to allow them to work in the usual way to prevent a strike. Further discussions with the loaders proved futile and the two researchers elected to approach Robert Sayre, Jr, the Assistant General Superintendent of the company. To their satisfaction, he empowered them to summarily dismiss anyone who refused to load iron on the piece-rate.

Next day, Gillespie and Wolle fired the entire gang. As a result, there being nobody left to observe, their experiment had ended before it had even begun. Nine days later, though, another foreman—John Enright—managed to 'persuade' some of his loaders to accept the experimenters' conditions. Again, there was not the slightest mention of scientific selection. These new volunteers were Henry Noll (in other words 'Schmidt'), John Strohl, Evan Miller, Preston Frick, Robert Skelly, Mike Morgan, and Tom McGovern. Importantly this entire group was either Pennsylvania Dutch or Irish. The Hungarian labourers, generally preferred as loaders, could no longer participate because their social leaders threatened to punish anyone of their ethnicity who co-operated with what they considered to be a pro-management policy.

Under no social obligation to the Hungarians, and needing the extra money, the Dutch and Irish were rather more malleable. Still, on the morning of 30 March 1899, only Noll, Strohl, Miller, Frick, and Skelly appeared for work. A day after starting the piece-rate, two more men dropped out. The five remaining spent 10 hours loading pig iron and managed to load 32 tons a man. This gave them a daily wage of $1.19 each, only marginally higher than they would have received doing day work. For those extra few cents, they were far more fatigued than usual and sloped home with sore and aching limbs. By now, Gillespie and Wolle were less than surprised to find that only three men, Noll, Frick, and Skelly, turned up for work the following day. Two days after the commencement of the experiment, more than half the team had dropped out. The normal way of working entailed some individuals moving the iron to the wagons where others carried out the stacking. The dire shortage of men demanded that entirely new methods of working be introduced.

Gillespie and Wolle instructed each man to load a separate cart without any help at all. This, no more than an unavoidable response to a labour shortage, seems to have been the extent of their training in new methods. As ever, Noll worked extremely hard and was soon loading the 45 tons per day to which Taylor refers. Skelly and Frick also loaded an impressive amount of iron but exhaustion soon took its toll. By 31 March, only Noll was still up to the job. To the temporary relief of Gillespie and Wolle, later that day two Hungarian brothers, John and Joseph Dodash, broke the embargo and volunteered for piecework. Yet, despite being physically fit, a week later they were still failing to make a fair day's wage and so left the gang. On 4 and 5 April, two more Hungarians followed the Dodashs' example and began to work on the piece-rate. Their act of defiance did not last long either. Within hours of coming to work on 5 April, they were relieved of their jobs exclaiming that 'their lives were in danger from the men who had been discharged from Haack's gang . . . [who] had threatened them with bodily harm if they worked by the piece'.

Once more, the experimental gang of pig-loaders had shrunk down to the trusty Henry Noll. It was therefore entirely appropriate for the final report to express sincere gratitude towards the 'little Pennsylvanian Dutchman' for being 'the one man who continued with us from the start . . . at times . . . constitut[ing] our whole piecework gang'.

What this makes abundantly clear is that Noll, like all the other men who had taken part in the experiment, had not been selected for his 'mental attitude'. The only criteria employed by Gillespie and Wolle in

forming their pig-iron gang was whether or not an individual was prepared to join it and, subsequently, whether that individual made a decent
wage doing piecework. To repeat, Taylor's colourful descriptions of the
key role personnel selection played in this are entirely fictional. From the
beginning to the end of the experiment, not a single man was turned away
by Gillespie and Wolle. They could not afford to refuse anyone.

During the second week of April, opposition to the piece-rate began
gradually to decline. First, the Hungarian workers were to some extent
pacified by Haack agreeing to rehire five of the Hungarians fired over a
month earlier by Gillespie and Wolle. Second, a new policy was introduced by which men exhausted by pig-iron loading on the piece-rate
were given less strenuous work until they were ready to return to loading;
this way they did not risk losing time and money in joining the experimental gang. As a result, by the middle of May 1899, Henry Noll had no
shortage of fellow loaders working the piece-rate. Some of these men
began loading as many as 70 tons a day, thus earning in excess of $2.60.
This level of achievement came as a real surprise to the researchers.
Finally, over the next couple of months, the experiment emerged as
something of a success. The general opposition to piecework subsided
and the other Bethlehem Iron Co. workers received their object lesson in
how both wages and productivity can be increased. Still, however,
Gillespie and Wolle continued to accept anyone into their gang who they
considered physically robust enough to take on this demanding job. The
intricacies of scientific selection were no more practised at the end than at
the start of the experiment.

But the account of scientific selection given in *The Principles of
Scientific Management* is not Taylor's only invention. Even his claims of
having overcome the hostility to piece-rates with the unintended help of
a local newspaper campaign is a concoction. Wrege and Perroni found
that the *South Bethlehem Globe* and the *Bethlehem Star* mentioned Taylor's
work at the Bethlehem Iron Co. only once between 1898 and 1901. The
single remark about Taylor is a flattering claim that the piece-rate system
would lead to an increase in employee wages. Aside from this, what was
going on in the field at the rear of the Bethlehem factory elicited no press
interest, much less the strongly negative campaign to which Taylor later
referred.

Such embellishments cannot be dismissed as the fruits of a failing memory drawn on after an interval of many years. Taylor's papers now enable us to plot the changes that accreted over time. In 1901, his remarks on these studies bore no relation to his famous 1911 account. Speaking about 2 years after the study had commenced, he told a meeting of the American Society of Mechanical Engineers that the only issue in selecting personnel had been physical ability. No reference was made to 'scientific selection' or 'careful training and instruction'. Only a 'properly built' man, he explained, could lift the 45 tons a day that Gillespie and Wolle considered reasonable. Embryonic versions of the account Taylor gives in *Principles of Scientific Management* only started to appear in papers dated 1903. They then progressively flowered into the full version. Perhaps to Taylor there was no great sin in this. He was convinced that his ideas were both sound and crucially important. A substantial gilding of the lily could perhaps be dismissed as being all in a good cause. But looked at from an historical perspective at least as cool and unyielding as Taylor's own, Wrege and Perroni demonstrate that there is more than enough material on which to dissent strongly from the conclusion Taylor reached about these studies in 1911:

> The reader will be thoroughly convinced that there is a science of handling pig iron, and further that this science amounts to so much that the man who is suited to handle pig iron cannot possibly understand it, nor even work in accordance with the laws of this science, without the help of those who are over him.

Given the total lack of evidence to support this claim, the modern expression 'dream on' now seems more than apposite.

5

THE HAWTHORNE STUDIES

Finding what you are looking for

> These investigations started out inauspiciously, as ordinary
> field work, consistent in most respects with the Taylor tradi-
> tion. They were just intended to be a bunch of straightforward
> studies of industrial hygiene factors . . . But a surprising series
> of events intruded on the theoretical background . . . the main
> point seems to be that the simple act of paying positive atten-
> tion to people has a great deal to do with productivity.
>
> Thomas Peters and Robert Waterman,
> *In Search of Excellence* (1982).

> But the results of subsequent experiments left the experiment-
> ers a little puzzled. After all their carefully controlled changes
> in hours of work, rest pauses and so on, they were unable to
> halt a general upward trend in rate of output. Even a lengthen-
> ing of the working day and a reduction of rest pauses seemed
> to have little or no depressing effect. The general upward
> trend, despite changes, was astonishing.
>
> Tom Lupton, *Management and the Social Sciences* (1970).

For more than 60 years the 'Hawthorne Experiment' enjoyed a
privileged place in books and courses related to human motivation,
the behaviour of small groups, and employee relations. In fact, few
who have studied psychology, sociology, or management science during
the past half-century could have avoided hearing of it. Unquestionably in
the first rank of social science research, it is famous for two reasons. First,
for 'scientifically' demonstrating that a humane, sympathetic management style can, on its own, lead to major improvements in productivity.
Second, for giving to the social sciences a brand new concept. According

Left: Fritz J. Roethlisberger (1898–1974).

to most accounts, the Hawthorne researchers were exploring, without preconceptions, how 'human factors' can affect productivity. When one of the groups being studied greatly improved its performance, they sought to determine why. The research team eventually located a surprising—almost magical—ingredient. Productivity had increased because of the interpersonal skills of an observer they had appointed simply to record what was going on. Even today, this kind of positive experimental contamination is still known as 'the Hawthorne Effect'.

In textbooks and lecture room, the Hawthorne story is frequently presented, as now, directly after a review of F. W. Taylor's ideas. To those viscerally opposed to all that Taylor represents, the Hawthorne findings came as manna from heaven. Taylor had characterized workers as people needing to be told precisely what to do and given a performance-related incentive before they would do it. In direct contradiction to this came the message that—treated with due sensitivity—workers were prepared to give of their best almost irrespective of the conditions under which they were working. The continuing appeal of the Hawthorne experiment flags up the great importance several fields still attach to this basic message. This is hardly surprising. The studies carried out at the Western Electric Co.'s Hawthorne Works in Cicero, Chicago, between 1927 and 1932 seemed to offer in themselves a total justification for the then emergent discipline of industrial psychology. Small wonder that they remain so popular with specialists in the successor fields of organizational psychology, social psychology, and human-resource management.

The bedrock of the Hawthorne edifice is *Management and the Worker* (1939), a book written by an industrial psychologist from Harvard University, Fritz J. Roethlisberger, and a member of Western Electric's management, William J. Dickson. As the authors make clear, the studies that began in 1927 followed on from earlier work grounded in Taylor's scientific management. In 1924, Western Electric, a subsidiary of the Bell Telephone Co., set up a joint project with the National Research Council of the US National Academy of Sciences to establish optimal levels of lighting for industrial workers. These were unsuccessful. First, when levels of illumination were increased 'production efficiencies by no means followed the magnitude or trend of the lighting intensities. The output bobbed up and down without direct relation to the amount of illumination'. Then, when lighting levels were cut back, production levels remained much as usual until the operatives could hardly see what they were doing. These findings brought out to the investigators 'very forcibly the necessity of controlling or eliminating the various additional factors which affect production output in either the same or opposing directions to that which we can ascribe to illumination'.

First Relay Assembly Group: the yardstick

Although not directly connected with the studies described by Roeth-lisberger and Dickson, the work on illumination provided the justifica-tion for a far more ambitious set of experiments ostensibly having the objective of teasing out just what those 'additional factors' might be. As productivity was going to be the yardstick against which success or failure was to be measured, the researchers first selected a job that was repetitive enough to provide reliable measures of output throughout. The task chosen was that of assembling telephone relays, each relay having about 35 component parts. To give greater experimental control, six 'girls' experienced in this work were persuaded to leave the main assembly room and work in a small unit some way from their colleagues. Two of the 'volunteers' were selected by the experimenters. The other four were chosen by the first two. Five were allocated assembly work, the sixth that of layout operator. The responsibilities of the latter included ensuring that sufficient components were always available and assigning work to the rest of the group. The layout operator also had minor supervisory respon-sibilities. Apart from this one operator, there was no direct supervision. But an observer was present in the test room throughout the experiment. In addition to making a record of all matters of possible relevance to the experiment, the observer's role entailed giving information to the group about each phase of the studies and seeking to elicit their views on what was going on. The output of each assembler was recorded electronically.

Having established a base level of output, the experimental group was then subjected to a variety of different conditions. An incentive scheme was introduced that was geared both to the smaller group and to indi-vidual effort. Morning and afternoon rest periods of 5 minutes were introduced and then lengthened to 10 minutes. These were then replaced by six 5-minute periods, which, in turn, gave way to single 15-minute morning and afternoon breaks, with a free meal being served in the former. The afternoon break was then reduced to 10 minutes. Next, the group was allowed to go home half an hour earlier and then an hour earlier. The normal finishing time and two 15-minute breaks were subsequently re-instated. Soon after they were given Saturday morning off. Following this, the morning and afternoon breaks were dropped and a full 48-hour working week was re-established. Finally this regime was modified by reverting to morning and afternoon breaks of 15 and 10 minutes, respectively. The free meal was not re-introduced.

Most secondary accounts report that throughout these changes production went relentlessly up. No matter what was done, or undone, the assemblers churned out more and more relays. Yet, Roethlisberger and Dickson's book makes clear that matters were not so straightforward. For example, the disruption caused by the six 5-minute breaks led to a drop in productivity. Similarly the loss of a full hour each day caused a predictable drop in output. Nonetheless, Roethlisberger and Dickson are at pains to point out that the general trend was up and that even when the original conditions had been re-instated at the end of the study, average output held at about 30 per cent above the opening position. This figure of 30 per cent became the baseline against which all other studies were judged.

In seeking to explain this very impressive rise, Roethlisberger and Dickson came up with five possibilities. First, that some physical characteristic of the changed working environment had been responsible. By this they meant, for example, improved lighting or ventilation. But this possibility was quickly dismissed because (1) no significant changes had been made and (2) the earlier lighting experiments had suggested that this was an inconsequential factor. Second, that the introduction of rest-breaks overcame a fatigue problem so effectively that, in the short run at least, the workers were able to cope with a reversion to working without breaks. After further investigation, Roethlisberger and Dickson rejected this hypothesis too. They did so on the grounds that medical evidence showed the assemblers to be working well within their physical capacity. The third possibility considered was that the rest-breaks resolved a monotony problem that resulted from long hours of highly repetitive work. This they partly dismissed on the basis that there was no evidence of boredom or apathy in the observer's records. But it was judged appropriate to investigate it further as a facet of supervisory style, their fifth hypothesis. The fourth possibility they put forward was that the 30 per cent increase arose directly from the improvements they had made to the employee incentive scheme. The fifth hypothesis was that the sensitive manner in which the team had been supervised throughout the study, particularly by the observer, had been the crucial factor.

Second Relay Assembly Group and the Mica Splitters

Obviously none of these factors were mutually exclusive. But Roethlisberger and Dickson soon concluded—albeit spuriously—that it was a two-horse race between the cash-incentive system and sensitive management. So they now faced the problem of determining the relative contributions of each. The strategy they adopted was twofold. First, they set up a second test group, this time on a shared bench in the main assembly room. Treated in all other respects the same as the other one hundred or so assemblers in the department, these operatives were taken off the departmental group incentive scheme and placed on the same incentive scheme as the first test group. Thus, the Second Relay Assembly Group had the same incentives as the group that had increased productivity by 30 per cent, but it had the same management style as the ordinary assembler. This was, in other words, a straightforward attempt to isolate the effect of performance-related pay (PRP).

This second study lasted for only 9 weeks. The tensions it generated in the main working group led to its premature termination. But it is not clear from *Management and the Worker* what caused these tensions. All the book speaks about is other workers wanting 'similar consideration'. We are left to wonder whether this was simply jealousy at the heightened interest shown; resentment that the new study group earned more money; or a concern that the greater output of the study groups would be used to tighten up the departmental incentive scheme. In any event, although the second study group increased their hourly output by an average of 12.6 per cent, Roethlisberger and Dickson went to some lengths in insisting that this was achieved very rapidly and thereafter seemed to plateau.

Their next approach was to set up another study group, this time comprising Mica Splitters, a class of workers who were already on a highly individualized incentive scheme. They were removed into a special test room and began to experience the full effects of 'friendly supervision'. No change was made to the PRP scheme on which they had always worked, so any increase in output could be ascribed to the changed managerial style alone. In the event, this group failed to match the approximately 30 per cent by which the First Relay Assembly Group had raised their output. The best the Mica Splitters achieved was about 15 per cent, a figure that fell away during the second year of the experiment as it became clear to the workers that their department was to be closed down and its work transferred out to California.

In seeking to explain the 15 per cent difference between the greatest improvements achieved by the Mica Splitters and those of the First Relay Assembly Group, Roethlisberger and Dickson make several suggestions. At one stage, they speculated that because the work of the Mica workers was so individualized, they lacked the incentive to work co-operatively in ways that had enabled the assemblers to make such progress. In other words, the human factors of increasing productivity had been under-estimated. But they finally contented themselves with the 'very tentative' conclusion that because this 15 per cent rise could not be explained as an effect of the performance-related pay scheme, it could be entirely credited to factors such as improvements in working conditions and supervision. This 15 per cent could then be deducted from the 30 per cent improvement achieved by the first group of assemblers on the basis that the remaining 15 per cent had to be the most that could be attributed to the new incentive scheme under which the first relay assemblers had worked. Although hedged about with some cautionary words, their over-all conclusion was that two separate studies had shown that a productivity improvement of at least 15 per cent could be achieved by the effects of creating a small, tight-knit team under a managerial regime sensitive to the workers' social needs.

Over the ensuing years the cautionary words have been forgotten. So, too, has the estimated 50:50 split between the contribution of the incentive scheme and that of the supervisory style. In the annals of the social sciences, Hawthorne is seen in relation to F. W. Taylor, as is Waterloo to Napoleon. What used to be known as 'rational-economic man' lies shattered on the ground; the victor, 'social man', goes forth to bigger things.

'. . . as them that won't see'

The question now to be addressed is whether the work described above is really fit to bear the edifice built on it. In doing so I have no qualms about judging the social sciences in accordance with the same principles as are applied to the natural sciences. This is exactly the standard set for them by founding fathers such as Auguste Comte, John Stuart Mill, Emile Durkheim, and Claude Lévi-Strauss, and this is a tradition still very much part of the identity of many modern social science faculties. In addition, the reverence with which the Hawthorne Experiment has been treated over the decades also serves to justify a critical examination. Although some modern texts now carry caveats, many (if not most) treatments of the Hawthorne Experiment uncritically offer up the original researchers' evaluations. It is a story that always finds a ready market. The audience wants to be told that they as individuals comprise far more than the crudely avaricious mechanoids implied by the Scientific Management approach. The social scientist, whose whole career is embedded in an entirely different paradigm, is more than happy to deliver. In this context, who wants to hear about sloppy methodology, poor observation, or misrepresentation? Instead, the famous pairing of Roethlisberger and Dickson is still frequently presented as having inaugurated a brilliant new conceptual approach.

But if, for many, it remains virtually unthinkable that Roethlisberger and Dickson's work should be omitted from general accounts of management and motivation, the difficulty is that careful reading of their own book suggests very strongly that it ought to be. In earlier chapters it has often been the case that the defects in experimental technique have been trawled up from private papers. Not so here. Every criticism raised below draws directly on the pages of *Management and the Worker* itself. It is illuminating, to say the least, that the manifold defects of the Hawthorne studies have only fairly recently been brought to the fore. This suggests that very few of those who have placed such reliance on the book have actually read the original with anything like a critical eye.

Changing horses mid-race

Unquestionably the most outrageous feature of the Hawthorne studies is the fact that two of the assemblers in the First Relay Assembly Group (that with the 30 per cent output increase) were replaced just under halfway through the experiment—that is, 40 per cent of the personnel were changed during the study. To the credit of the original authors, no attempt whatsoever was made to conceal this in their book; in contrast, a myriad secondary sources simply expunge what happened from the record. Roethlisberger and Dickson's offence lies in failing to address the major issues the event raised.

The facts as presented in the book are these. The criteria for selecting or accepting the original team members was that they be 'thoroughly experienced' and 'willing and co-operative'. In other words, the researchers isolated a group of women always likely to be more productive than the average operative. As it happened, this selectivity was not initially of much importance because the experimental conditions seemed to have little positive effect. For the first few weeks the test-room observer worked hard to create and maintain a friendly atmosphere. He chatted affably with the assemblers during work hours and, outside of them, parties were organized.

Despite this, only a very minor increase in productivity was achieved. The unwelcome message seemed to be that sympathetic management was not significantly enhancing productivity. During the next 7 months the researchers learned what many of their more cynical colleagues would have predicted from the start. Having been told at the beginning of the study to 'work as you feel', some of the operatives interpreted their instructions as a licence to do as little as they wanted. Which wasn't much at all. Women initially recommended for high productivity began to spend a gratuitous amount of time chatting and relaxing during work hours. Assemblers 1A and 2A were seen as the prime culprits, to the extent that Roethlisberger and Dickson later accused them of having 'jeopardized' the entire experiment. But at no time did Roethlisberger and Dickson consider the possibility that their behaviour might have been encouraged by a permissive management style.

After weeks of poor performance, firm measures were finally taken.

The operatives were forbidden to talk and threatened that if they did not work harder they would lose their free lunches. Even this appeal to their stomachs—ironically working on the classical assumption that motivation is largely a matter of the satisfaction of basic physical wants—had little effect. Over the next few weeks, 1A and 2A were repeatedly told off and threatened for being 'moody', 'inattentive', 'unco-operative', and having inappropriate 'mental attitudes'. Roethlisberger and Dickson's Harvard mentor, Elton Mayo, subsequently claimed that, 'At no time in the [study] did the girls feel that they were working under pressure'. Roethlisberger and Dickson give the lie to this. They relate how operator 2A was called into a 'conference with the test room authorities' and put under pressure to repent and correct her waywardness. Despite being 'apologetic and promis[ing] to improve' when questioned in this way, her 'old attitude returned immediately'. As a result, in 'the best interests of the study' 2A and 'her ally' 1A were summarily dismissed from the test room for 'gross insubordination' 4 months into the experiment.

Far from being sympathetic and friendly, the test-room management had become intolerant of any signs of lowered productivity. Within a few weeks of the study beginning they had ceased to be the dispassionate observers of the standard textbook account. And even though it later proved possible to revert to more friendly ways, there can be little doubt that what had happened to 1A and 2A was recalled with some trepidation by the other team members; Voltaire's famous phrase about the English occasionally shooting an Admiral 'in order to encourage the others' springs to mind. Roethlisberger and Dickson justified the removal of 1A and 2A on the grounds that if they were to test the effects of friendly management on the women's performance, then they needed to be able to:

> Treat the girls' attitude towards the test as a constant factor; therefore the 'right' mental attitude was essential. [Their] attitude towards the test was something the girls themselves could and should control. . . . [As a result,] for any failure in this respect the girls were held responsible.

This is a remarkable passage because if the operatives were seen to be taking advantage of the study conditions—which they almost certainly were—this information was clearly germane to any evaluation of the utility of friendly management. At one point in their book, Roeth-lisberger and Dickson conceded that the investigators ought to have inquired 'into the causes of the [2A's] problem'. This they eventually did, but in no way did they let this pivotal episode affect their interpretation of the results.

At the very least, the study should have been abandoned at this stage and restarted with new team members. Though, arguably, the very public drama of the two personnel being removed had so contaminated the experimental environment as to make even a re-start a scientific non-sense. Assemblers were keen to join the experimental set-up because it offered congenial working conditions and various combinations of free meals, rest periods, and shorter working hours. But after two assemblers had been chucked out for slacking, they began to feel that quid pro quos were involved: work hard and stay, or slack off and get out. But it did not suit the researchers to think along these lines. Instead, their response was to argue that the behaviours of 1A and 2A were somehow pathological and therefore unrelated to the test conditions. Accordingly, one researcher described 2A as having 'paranoid preoccupations, fatigue and organic dis-ability'. Yet if this had been so why would she have been chosen in the first place? These characteristics (if they existed at all) became manifest only after she had walked from the factory floor into the test room itself.

Further problems arise from the way in which their replacements were selected. This is Roethlisberger and Dickson's account of what happened next:

> The foreman, who chose the new girls, was asked to select girls who were experienced relay assemblers and desirous of participating in the test (these were the requirements used in selecting the original operators), and, moreover, whose hourly rates and weekly output performance were comparable to those of the operators to be replaced. These additional requirements were necessary in order to avoid altering the distribution of earnings among different members of the group and also to avoid too great an interference in the output data.

This is to claim that the replacements were selected so as to minimize the discontinuities from the previous stage of the experiment. But this is clearly fatuous. The earlier operatives 1A and 2A had been fired for 'gross insubordination'. Putting two dedicated workers in their positions was hardly replacing like with like. Indeed, just how different these replacements were becomes clear if we delve a little deeper into the attitude and circumstances of the new operative 2.

Once again, Roethlisberger and Dickson make the relevant facts known to us. When operative 2 joined the group, her father had recently lost his job and she had become the family's sole breadwinner. Poverty and duty combined to make her want to take maximum advantage of the performance-related pay scheme operating in the test room (exactly the sort of credentials Taylor would have drooled over). Operative 2's determination to increase output, combined with her forceful personality, soon made her the informal leader of the group; and the enthusiasm of both new operators was shown in their immediate achievement of levels of productivity far exceeding any of the original team.

Yet, such performance differentials were a source of real frustration to operative 2. The wage-incentive scheme paid according to the average output of the entire group, so she stood to lose out because of the inefficiency of other group members. She was not about to let this happen. When the output curves showed that the productivity of operators 3, 4, and 5 was on a downward trend, her immediate outburst revealed how much the extra money meant to her and her family, 'Oh! What's the matter with those other girls. I'll kill them.' Throughout the study, operator 2's private worries drove her to work harder than anyone else. And it was also her dedication that inspired the new operative 1 to work unusually hard. Sympathetic supervision was of no obvious relevance to either of them.

Unsurprisingly given this backdrop, the 30 per cent increase in productivity of which the researchers later boasted was mostly the achievement of replacement workers 1 and 2. This is crucial because the study records prove that their output was already very high *before* they could have benefited from the new supervisory regime. Seeing the main chance, assembler 2 in particular 'hit the deck running'. Yet in the study's most egregious confusion of cause and effect, Roethlisberger and Dickson were later to claim that high output and relaxed, sympathetic management coincided in the last months of the study because the new management–staff relationship had won the loyalty and confidence of the operators. Almost certainly the reverse is true. Once the new operatives began to raise productivity, the researchers were content to relax the aggressive management style they had previously been forced to adopt. In other words, the increased output caused the return to a 'friendly' management style.

Seeking salvation

Although not quite of the same magnitude, there were other serious errors in the Hawthorne Experiment. It is clear from the way in which the rival hypotheses for increased productivity were dismissed that the researchers had an agenda. For example, one worker in the First Relay Assembly Group was given the task of assembling all non-standard relays allocated to the test room. Of the five, her productivity improvement was the lowest. This strongly suggests that simplifying work tasks had had a generally beneficial effect on productivity. Yet Roethlisberger and Dickson simply wrote off this evidence as 'inconclusive' without making any attempt to explore it further. Knowing what they wanted to find, they moved on with extraordinary equanimity in its pursuit.

Next, the researchers examined the potential effects of fatigue on productivity. As we have seen, they soon found evidence of a positive correlation between shorter working hours and higher hourly output. Reduce the hours or give longer breaks and increase the productivity. Yet, this finding also conflicted with their prior conviction that social factors alone were capable of having a major impact on productivity. So Roethlisberger and Dickson asserted that because there had been no 'medical evidence' of fatigue among the workers, fatigue had not been a problem. Quite apart from the fact that no attempt was made to isolate this factor and explore it in more detail, their rationalization that fatigue has to be at so gross a level that it can be medically certified makes little sense.

Roethlisberger and Dickson also adopted the stratagem, when discussing both task complexity and the effects of fatigue, of rejecting inconvenient explanations on the grounds that that particular factor, on its own, could not offer a sufficient explanation for the 30 per cent increase in productivity. Significantly they did not subject the supervisory relationship to this 'all-or-nothing' test. It is inescapable that throughout the study, Roethlisberger and Dickson were determined that sympathetic supervision and positive group responses should come out on top. But their cavalier sweeping away of possibly significant contributory factors is nothing in comparison with their treatment of the much-improved incentive scheme. Roethlisberger and Dickson clearly felt this to be the main threat to their belief in the primacy of social factors. For this reason they were determined that the credit PRP received was very strictly limited. To achieve this they used techniques that would bring a blush to the face of a riverboat cardsharp.

It has first to be appreciated that the researchers heavily compromised their own agenda both in getting the First Relay Assembly Room established and in keeping it running when the behaviours of operators 1A and 2A so seriously threatened it. The initial problem was to get assemblers who were already working under a remunerative incentive scheme to participate in the experiment. Had Roethlisberger and Dickson wished rigorously to test the effects on productivity of a benign supervisory regime, there was an incredibly straightforward way of doing so. They would pay the assemblers a guaranteed wage that matched their normal earnings and then see what good relationships on their own could achieve. Astonishingly this does not even seem to have been contemplated. Instead, Roethlisberger and Dickson devised a much more effective incentive scheme, which, instead of being based on the combined efforts of the whole Assembly Department, was tightly focused on the six workers in the test room. Then, when in spite of this enhanced incentive and the initially benign supervision, two workers performed so badly that it was found necessary to replace them, their chosen replacements were a work-study engineer's dream: two girls who, from some combination of practical and emotional factors, were hell-bent on making as much money as they could.

It cannot have been too long before those running the experiment realized the sticky mess they were in. Although, superficially, the replacement assemblers seemed to have saved the day, the conversations recorded by the Observer made it all too clear why they were working so hard. Desperate remedial action was clearly needed; hence the setting up of the Second Relay Assembly Group. But, as has been seen, here things went from bad to worse. Because this was an attempt to isolate the effect of the PRP scheme in the First Relay Assembly Group, the new group was treated in almost all respects like the assemblers in the main department with whom they remained. The sole changes were that they were physically grouped together and were paid in accordance with the incentive scheme applied to the group in the first test room. No commitment to 'friendly management' was made. Naturally the experimenters were hoping that any rise in productivity would be very small or non-existent. This would allow them to claim that the 30 per cent productivity rise in the first group's test room could be accounted for only by the effects of the supervisory regime. But things did not work out that way. As we saw earlier, once the small group-incentive scheme was introduced, the second group almost immediately showed a 12.6 per cent increase in hourly productivity, an increase it had taken the first group 9 months to achieve. One possible inference that could be drawn from this is that if sympathetic management had had any effect at all, it was entirely negative.

Again as seen earlier, this experiment was soon abandoned on the grounds that it caused too much tension amongst the other assemblers. Nor does this seem unlikely. Along with more productivity came more pay and more attention, both likely to be a potent source of jealousy. We can also imagine that the departmental conspiracy theorists would have gone into overdrive as the new team's productivity rose. Using a term the researchers later picked up from another group they studied at Hawthorne, the five girls would have been accused of 'rate-busting'—in other words, giving the management grounds for demanding more work for the same money in the departmental incentive scheme.

These factors alone could well have caused the Hawthorne management to terminate the experiment. What we can now infer, however, is that this did not come a moment too soon for the experimenters. Had it continued there was every danger that they would have had to publish a report vindicating Taylor's carrot-and-stick approach. Yet, at no point in *Management and the Worker* is credit given to the possible potency of the incentive scheme. Instead, the Second Relay Assembly Group's rapid increase in productivity is put down to the social effects of working in a small group and, in particular, intergroup competition. Thereafter, the

12.6 per cent increase the second group achieved in 9 weeks is treated as directly comparable with 'the roughly 30 per cent' achieved in the test room in about 2 years. To use a term that would be familiar to our river-boat cardsharp, the new incentive scheme was up against a stacked deck.

Clearly, though, the Second Relay Assembly Group had not got the researchers out of trouble. In fact, it had done the reverse. In the face of this awkward finding, their next step showed—if nothing else—both resilience and creativeness. In recruiting Mica Splitters for their third study, they were drawing on a group that was already on an individual incentive scheme. This meant that performance-related pay could not be responsible for any increase in productivity. Within this group substance was given to the 'enlightened supervisory regime' by simplifying work, increasing rest periods, and shortening the length of the working day. Because the direct impact of such factors on productivity had been dis-counted in the First Relay Assembly Room, however, the researchers felt able to assign any increase in productivity to friendly supervision alone.

In the event, as seen earlier, an increase of 15.6 per cent was recorded. The researchers were delighted. But they ignored the fact that other conditions almost certainly made a major contribution to this total. In particular, although they admitted that the study was being conducted in a rapidly worsening economic climate, they argued that this could explain the marked drop in productivity only during the second year. Certainly by the second year it had become obvious to the Mica Splitters that their jobs were to be transferred to California. But what must the effects of this have been when the mass redundancies were still only a possibility?

It is probable that during the first year, when the 15.6 per cent improvement was achieved, the workers in the group under study believed that prodigious displays of hard work might yet save them from the maws of the Great Depression. This is consistent with the general upturn in productivity in the rest of the factory during this period. It also accords with remarks, documented in the book, made by the Mica Splitters towards the end of their most productive period:

> **Operator M3**: Right now I am very satisfied. I couldn't ask for any-thing better than I have now; in fact I would be very unreasonable if I did. I hope I am kept on in the test room as long as I am working at the Western. . . . I love this little test room.

Operator M4: I never dreamed it was going to be as good as it is, but when I was asked I thought I would take a chance. . . . There were three others asked but they foolishly didn't take it. . . . Now that the work is dropping off, every time the other girls meet us, they ask 'Are you still on Mica?' When we tell them we are, I can see they are sorry although they don't say it.

Read more than 70 years after the event, the desperation in these words seems to leap from the page. But to the researchers, these and similar comments are naively seen as no more than a vindication of their preferred management style. We should also remember that once the operators had delivered what the researchers wanted, sympathetic management style notwithstanding, the five Mica Splitters would almost certainly have shared the same dismal fate as their colleagues.

Difficulties with figures

Roethlisberger and Dickson, having attained their prize, were not above gilding it. Their strategy is clear. They needed to be able to point to a considerable output improvement free of the taint of the cash incentive. So, from their point of view, the greater the improvement in output in the Mica Splitting room the better. Ever since the publication of *Management and the Worker*, readers have not unreasonably assumed that the 15.6 per cent referred to above is the difference in monthly productivity between the operatives' first and last months in the Mica Splitting test room. A perfectly natural assumption, but quite wrong. This figure was actually derived by averaging the peaks in the hourly productivity of the five women. At no point was the entire team working with 15.6 per cent greater effectiveness: for most of the time they were working at a substantially lower rate, much closer to the output level typical elsewhere in the factory. Indeed, because the length of the working week had been considerably shortened, in absolute terms output actually fell. The judicious choice of the maximum hourly figure served conveniently to mask what was otherwise a rather embarrassing finding.

Skipping over these awkward inconsistencies, Roethlisberger and Dickson then subtracted the Mica Splitter's 15.6 per cent from the First Relay Assembler Group's 30 per cent and boldly concluded that no more than 15 per cent of the growth in output achieved by the latter could be attributed to the incentive scheme. Then, leaping over all other possible variables, they went on to suggest that at least the remaining 15 per cent was due to the introduction of the better supervisory regime. Yet, their own figures better support an entirely opposing set of conclusions. Looked at more deeply, it is clear that the Mica Splitting study does not show that friendly supervision alone boosted total output. Quite apart from the fact that productivity hardly grew at all, wider economic circumstances were strongly encouraging harder work. Nor should the effects of the removal of 1A and 2A be forgotten. Furthermore, the Second Relay Assembly group had shown a very rapid increase in productivity seemingly as a direct result of the introduction of a wage-incentive scheme. It is therefore reasonable to ascribe almost the full 30 per cent increase achieved by the First Relay Assembly Group to the improved incentive scheme they worked under.

Had Roethlisberger and Dickson not been so hostile to the Taylorite approach, they could have published their findings as a compelling validation of his ideas. Even the productivity increases that should almost certainly have been attributed to rest breaks and shorter hours served to vindicate Taylor's highly physiological conception of labour. Add to this the facts that control groups were never used, that groups of five people are too small for proper statistical analysis, and that no attempt was made to ensure that the selected groups were representative of the general population, and the worthlessness of the Hawthorne studies becomes apparent.

By a policy of fudging, wishful thinking, and highly selective analysis, Roethlisberger and Dickson managed to ignore both the evidence for the effectiveness of wage incentives and the devastating flaws in their own experimental approach. This is not for a moment to suggest that the manifold shortcomings of the Hawthorne Experiment mean that the social side of work is either unimportant or non-existent. Instead, this account teaches us that Roethlisberger and Dickson's conclusions simply cannot be relied on to offer any valid insights into this exceptionally complex area.

Before closing this chapter, I wish briefly to consider one last theme. Roethlisberger and Dickson's clarity of expression and their ability to 'manipulate' data on such a grand scale indicate a high order of intelligence, conscious or otherwise. So what drove two exceptionally able men to produce what can fairly be described as a travesty of scientific research?

'A debate with Marx's ghost'

In the minds of many students—and perhaps also in some cases the minds of those teaching them—the original 'illumination studies' get merged into the story of the First Relay Assembly Room. From this emerges a picture of a company firmly wedded to the Taylorite approach, forced by an accidental discovery to come to terms with the full richness of human potential. As I have shown, Roethlisberger and Dickson cannot be blamed for this telescoping of events. They make clear that the illumination studies were an entirely separate exercise. What cannot be questioned, however, is the effort they put into presenting their findings as arising from an unexpected discovery. When referring to an early phase in the First Relay Assembly Room they claim, 'the significance of the differences between supervision in the test room and that in the regular department had not yet come to the explicit notice of the investigators'. Elsewhere, they account for a lack of scientific rigour in seeking to isolate the effects of supervisory style by explaining that for a considerable time its importance was not appreciated.

Yet, the opening section of *Management and the Worker* includes a footnote that calls this very much into question. It lists the 'Employee Relations Policies' to which all Western Electric managers and supervisors were required to adhere. Known as the 'Ten Commandments', the first eight stress the importance of providing good pay and conditions, continuous employment, work matched to abilities, career development and training, help in times of need, encouragement of saving, and opportunities for recreational activities. The last two warrant quoting in full:

> IX. *To accord to each employee the right to discuss freely with executives any matters concerning his or her welfare or the Company's interests.*
>
> It is your duty to establish the conviction among those whom you direct or with whom you come in contact that sympathetic and unprejudiced consideration will be given to any employee who wishes to discuss with you and with Company executives matters of his or her welfare or the Company's interest.

X. *To carry on the daily work in a spirit of friendliness.*

As the Company grows it must be more human—not less so. Discipline, standards, and precedents become more necessary with size, but the spirit in which they are administered must be friendly as well as just. Courtesy is as important within the organisation as in dealing with outsiders. Inefficiency and indifference cannot be tolerated, but the effort of every supervisor must be increasingly directed at building up in every department a loyal and enthusiastic interest in the Company's work.

There are two surprising things about these policies. First, they seem so modern in style and outlook. Second, they are dated 'May, 1924'. As this is over 3 years before the Hawthorne Experiment began, we are forced to choose between two possibilities. Either Western Electric had forgotten about two of its key personnel policies during those 3 years, or, more likely, the study's findings were not so unexpected after all. Given this background, it seems highly probable that the real objective of the studies was to demonstrate the productiveness of friendly supervision, and the only surprise encountered was the sheer difficulty of doing so.

On this reading, the fake surprise was simply a means of adding credibility. As every snake-oil salesman knows, having an apparent stranger say, 'Your chilblain cure did wonders for my baldness', sells far more snake-oil than does the simple cry, 'Buy my baldness cure'. If some attribute of the salesman's product seems to surprise the seller as much as the customer, it appears more likely that that attribute is real. This is not to suggest, however, that Roethlisberger and Dickson cooked their results quite as consciously as would out-and-out fraudsters. Both were convinced that relations between management and staff were key factors in promoting a smoothly running organization and therefore persuaded themselves to 'see' evidence when none in fact existed.

Roethlisberger and Dickson's Harvard mentor, Elton Mayo, had a pivotal role in precipitating this purblindness. Mayo was a passionate believer both in the rationality of capitalism and the need for managers to act in order to avert the triumph of Bolshevism. His worldview was that of a member of an educated elite that deeply valued its close ties with many of the great industrialists of the age. In the first half of the twentieth century, with little or no knowledge of the gulags, secret police, and mass starvation, communism seemed an attractive option to many Western workers and intellectuals. Karl Marx had predicted that conflict between the industrial work force and their bourgeoisie employers would inevitably lead to the overthrow of capitalism. To Mayo, Taylorism represented an unintended high road to this stage of worker disaffection and revolution. For this reason, he was striving for what others have called a 'third way', a means by which human potential could enrich capitalism rather than being crushed by it.

From the long perspective of history we can see that such grandiose ideas were never likely to catch on in the 1930s. The Great Depression afforded no time for sentiment. What became known as 'human relations' had to bide its time until the power relativities between capital and labour swung once again in favour of the latter. During the Second World War and subsequently during the period of full employment, labour shortages and rising trade-union power meant that, once again, being nice to people could come back into favour. Then, *Management and the Worker* was taken back down from the shelf and commenced its long years of ascendancy.

When the Hawthorne Experiment began in 1927, the economic conditions were broadly comparable to the post-Second World War boom years. But as they started to write up its findings, Roethlisberger and Dickson would have been acutely conscious of deteriorating relations between management and staff as the Depression deepened. They shared Mayo's implacable opposition to the Marxist belief that class warfare is endemic. Accordingly, they also attached enormous importance to Mayo's claim that humane supervision could open up whole new vistas of harmonious labour relations and, consequently, more efficient capitalistic production. This led them grossly to distort and misrepresent their experimental findings. Their intentions may well have been honourable. But at this distance, *Management and the Worker* serves primarily to reinforce our awareness of the supreme difficulty of excluding prior beliefs from studies that touch directly on social and political issues of great sensitivity.

Conclusion to Part one

Sins against science?

I f one thing characterizes all five cases in Part 1, it is the making of very bold claims on the basis of less than comprehensive evidence. If nothing else, this cautions the historian against using such limited evidence to mount a general indictment of the way in which the scientific community conducts its business. I chose the cases because they show how wide the gap between myth and reality can be: they in no sense constitute a representative sample. As of yet, the history of science is too young a discipline to have built the kind of database necessary for us to form a balanced view as to how typical such behaviour really is. All we can say now is that a not-inconsiderable proportion of the scientific greats examined by historians have had a real world existence at considerable variance with the near-perfect characters attributed to them by myth.

Nevertheless, although this may be a discrepancy from which only the great suffer, intuitively, this seems unlikely. Common experience tells us that feet of clay are not unique to heroes. In all probability, manipulation of experimental data is not just the sin of a few great men who have somehow managed to slip through the net. Their particular 'warts' have been exposed for no other reason than that their fame has attracted historians to them. And although I wouldn't for a moment suggest that the sorts of chicanery practised by F. W. Taylor even approach endemic proportions, it seems reasonable to hypothesize that as pressure on research topics drives future historians to lower levels in the scientific-merit order, comparable examples will not be wanting. The only difference to be expected is one of scale: the temptations to which Pasteur, Eddington, Millikan, Taylor, Roethlisberger, and Dickson were exposed were enormous. Consequently the risks they were prepared to take—consciously or unconsciously—were concomitantly large. In more low-key areas of scientific enterprise the sins would be scaled down. Yet they would probably be just as real and, to some at least, just as tempting.

If we are to draw any general lessons, we can make a start by using our chosen examples to consider what sorts of behaviour are seriously unbecoming of a scientist. Put another way, just how badly did the scientists we have looked at behave? To those strongly committed to the traditional view of the scientific method, their most obvious shortcoming will be their inability to have stood aside from their preconceptions. We know how the textbooks say good science is carried out: great men of the past from Francis Bacon, through Isaac Newton to Charles Darwin have all inveighed against the deductive mode of reasoning. They have insisted that rather than working from the generality of theory to the particulars of fact, the scientist must first collect the data from which reliable theories will naturally arise. This is why, in his autobiography, Darwin claimed that he had followed 'true Baconian principles'. By this he meant that it was only after he had first 'collected facts on a wholesale scale' that he turned his mind to the construction of explanatory theories. The pivotal importance of this approach has been so firmly believed in that philosophers used to think science could only progress at a rate dictated by the accumulation of relevant data.

The difficulty now is that this model is clearly wrong. Most scientists begin with hypotheses derived from very limited evidence. If for no other reason than economy of effort, theories usually precede the large-scale accumulation of facts. And, for all his protestations, Darwin was no exception to this rule. Several historians have now shown that from the moment he began his evolutionist speculations, he was strongly guided by one or other of the pre-existing theories of how species change over time. Without the aid of such ideas, Darwin would soon have choked on the volume of 'facts' he had managed to collect within weeks of embarking on his grand project. As this suggests, for all the dangers involved in having a-priori beliefs about the way in which nature works, such assumptions are indispensable if sense is to be made of data derived from even the most apparently precise of experiments. Experimental or field evidence nearly always contains some degree of ambiguity. Order must therefore be imposed on it. If we did not do this, progress would at best be tortuously slow; at worst, impossible.

There is also an inevitability about preconceptions at a deeper level. The role evolution seems to have assigned the human population is that of problem solver and opportunist. Our modus operandi is to look at novel situations in terms of the potential they offer for exploitation. This means looking for causal relationships and underlying principles. To find these we draw on our experience. We do not, therefore, respond randomly to new situations, but on the basis of the ideas and preconceptions we have already formed. In the event, these may prove to be in urgent need of modification; but this does not mean that our approach is fundamentally flawed. Provided we have proceeded with caution and immediately recognize when an established strategy is failing, we have our preconceptions to thank for providing an initial basis for action. An organism that felt the need to tackle every new event from first principles would be unlikely to survive, let alone multiply: there is simply too much competition and too little time for unconstrained enquiry to be the evolved approach. So, given that preconception as an essential guide to action has seen us through the fire of natural selection, it would seem unreasonable to think that once we embarked on the great scientific enterprise, it should have been entirely abandoned as an operating principle. The great error lies not in starting off with an idea, but in clinging on to it against all the evidence.

What has been said so far suggests that the much-vaunted term 'the scientific method' may need some redefinition. Most would agree that at its broadest the expression denotes research conducted exclusively in accordance with the dictates of rationality. But delineating what pure reason does and does not involve in the conduct of scientific research is far from easy. If preconception is an essential component of rationally conceived scientific investigation, according to many philosophers and scientists so are personal ambition, the periodic exercise of authority, and the suppression of awkward or inexplicable results. Personal ambition is the motor that drives most scientists to come up with new ideas and persist with them even if the first signs are not entirely positive. Pulling rank can be a necessary means of ensuring that debate isn't derailed by those without the knowledge and experience to offer the best of judgements. And discarding some results can be justified where new and unpredictable technologies are being employed.

As the conventional model of good science would exclude each of these as unworthy of the scientist, we need to replace it with a model that better accords with reality. Taking such a view enables us to see that some of the scientists I have described were deviating not from the path of reason, but from an idealistic notion of good science that is neither realistic nor necessarily based on rationality; the flaws sometimes lay in our standard definitions of the scientific method rather than in what the scientists actually did. So, if it's not possible to define precisely what good science is, we must be prepared to judge each case on its merits. To do this we need to adopt an inclusive definition of appropriate scientific behaviour and constantly bear in mind that human factors are as likely to accelerate as to impede scientific advance. In addition, lest we fall into the trap of presentism we must be sure that our judgements in no way conflict with those that would have been made at the time.

Even having made these concessions, I think we would all agree that F.W. Taylor committed serious sins against science. There is no doubt that he had a burning conviction that skilled industrial engineers could identify 'the one best way' of doing a job that should replace inferior methodologies thrown up by custom and practice. He also believed that skilled personnel selection would provide a better match between worker and task than could the undiscriminating vagaries of the job market. Likewise, he never questioned that a well-designed incentive system could greatly improve productivity. These are all important ideas that in some conditions have been shown to bear out the claims he made for them. But very little of what actually went on at the Bethlehem Iron Co. can properly be adduced in their favour. In science, for the entire period covered by this book, the wholesale distortion of the experimental record has been deemed thoroughly unacceptable. Taylor clearly failed to meet this most basic of standards.

Arthur Eddington is a case in which students of science are brought hard up against Bismarck's unconscionable dictum 'We should never allow our principles to get in the way of our opportunities'. The facts now seem clear enough. Neither of the teams sent out to measure the deflection of light during the 1919 solar eclipse was adequately equipped to make definitive measurements in the field. Unsurprisingly the results they obtained were inconclusive and widely scattered. Some of the best seemed to favour the Newtonian view rather than general relativity. Eddington knew precisely what he wanted, however, and selected or rejected the results in strict accordance with one principle: whether or not it supported Einstein's theory. Those data-sets that did, were in; those that did not, were out. But, rather than castigate Eddington, there are those who would raise a powerful case for mitigation. If we accept the argument that Newton was the last of Einstein's precursors to have had ideas of equivalent importance, we are talking about contributions of extreme rarity. Given that to gain acceptance they will have had to over-turn paradigms of exceptional robustness and generality, it can be argued that anything that enables such radical ideas to get a head start does nothing but good for science.

For the sake of argument, let's assume that Eddington's efforts gave Einstein's ideas a boost that brought them into the mainstream 5 years earlier than would otherwise have been the case. Does this outweigh the means by which the head start was obtained? As this question is routinely put: do the ends justify the means? My judgement is that in this case at least, they don't. An impartial scientific observer of 1919 would almost certainly have agreed. No doubt the prize was great; but buttressing woe-fully inadequate data with personal prestige, power, and influence is too high a price to pay for it. Einstein's ideas were strong enough to have made their own way in the world: they did not require help that itself serves to debase the whole scientific enterprise.

A similar plea in mitigation can be entered in respect of Fritz Roethlisberger and William Dickson. What, after all, was their sin other than to seek relentlessly for evidence proving that it pays to be nice to people? To this, too, I think we should turn a hard face. As with Taylor's ideas, there may well be circumstances in which the 'human relations' approach pays handsome financial dividends; but to treat such a proposi-tion as a scientific fact would require far better and far more nuanced work than Roethlisberger and Dickson produced. Indeed, it can be argued that their efforts were counter-productive to their own ends.

The ultimate justification for treating people decently is ethical. Those accepting this as a cardinal rule aspire to stick with it even when the cost-benefit ratio militates against it. To use a travesty of good science as a basis on which to treat such a principled position as no more than intelli-gent self-interest, undermines the objective being sought. During periods of staff shortage, a strongly employee-centred approach may well prove economically rewarding. Unfortunately at some stage, an economic downturn—even if not on the global scale experienced at the Hawthorne Works—will dramatically change the key inputs to the equation. As the bottom line shifts into the red, it is far more comforting for managers to see harsh measures as a timely corrective to bad science, rather than as the abandonment of a fundamental principle. For the worker on the receiving end of a harsh and bewildering change in personnel policies this may well mean that hard times come sooner rather than later. Add to this the abysmally poor experimental controls of Roethlisberger and Dickson's study, and the fundamental dishonesty of their presentation, and it again seems to me that a harsh judgement is amply justified.

We are now left with Louis Pasteur and Robert Millikan. Here the evidence is both more equivocal and more instructive. Although it is an axiom of science that no theory can ever be termed indisputably correct, posterity has judged that both men backed the right horses. Pasteur's work led on to the germ theory central to modern medicine and the vanquishing of the idea that the spontaneous generation of life is a common occurrence; Millikan built up an understanding of the electron that continues to inform modern physics. The charge against them is that they failed to disclose all the information they had at their disposal and, in Pasteur's case, that he wilfully failed to follow up lines of inquiry that gave encouragement to his opponents.

We can sympathize with both men. With the invaluable aid of hindsight, we can see the terrible pitfall that lay in wait for Pasteur. Because the whole of the scientific community mistakenly was locked on to the notion that even microbial life forms cannot tolerate any significant exposure to boiling water, easily adduced evidence to the contrary would have been misinterpreted as an incontrovertible demonstration of spontaneous generation. Pasteur's career would have been tarnished and the emergence of the science of microbiology set back for years; and he would have been proven wrong for all the wrong reasons. Unlike Eddington's attempts at experimental proof, most of what Pasteur did was brilliantly conceived and conducted with consummate skill. Given that these outstanding efforts provided Pasteur with ample grounds for believing in the overall strength of the case he was making, I think we may say that his sins were of a comparatively modest nature.

So, too, it seems to me with Millikan. The naive honesty of the first published account he gave of his work was exemplary in both senses of the word. Although the degree of openness he displayed set a standard others might have been expected to follow; the extent to which his own words were turned against him shows why, in practice, they would have been ill-advised to do so. His fingers badly burned, by the time he next came to publish Millikan was careful to suppress any uncomfortable data. How he justified this to himself we do not know. But one possibility, frequently discussed by philosophers of science, is that he was earnestly trying to protect a potentially important idea from the premature savaging it would receive were he totally honest about his experimental results.

It takes a long time for new experimental apparatus to be made entirely reliable. But Millikan's difficulties were even greater than most. Trying to determine the charge of individual electrons from the behaviour of comparatively huge droplets of oil was always going to throw up some rogue results. Fully aware of this, Millikan must have realized that because the electron theory was as fragile as a newborn child, such results could easily compromise its survival. Indeed, such is the savagery of scientific debate that we, too, must accept that there was little chance of the subtleties and limitations of his experimental method being taken into account by his detractors. A host of uncharitable peers was waiting in the wings to proclaim his evidence inadequate and his theories miles wide of the mark. Perhaps, then, Millikan came to see that he had to buy the electron theory some time by suppressing his more 'awkward' data-sets. Within a few months, he may have reasoned, his experiment would be rendered all but infallible and he would be able to reflect on his modest piece of deception as a gamble worth having taken. Instead of swimming against a tide of opposition and scepticism, he would then be calmly consolidating a victory already in the bag.

Looked at from this point of view we can make a good case for Millikan being seen as less guilty than the system itself. He was operating within a scientific culture in which ideas can be smothered at birth. Why? Because, as a result of egos becoming strongly attached to ideas, there is a profound vested interest in ignoring the true difficulties involved in proving any new theory that requires the use of innovative experimental apparatus. The Nobel Prize-winning physicist-philosopher Percy Williams

Bridgman once remarked, 'The Scientific Method is doing your damnedest, no holds barred'. Yet such is the adversarial nature of scientific competition that 'no holds barred' has acquired too literal a meaning. So, until scientists feel able to publish ambiguous data without this evidence of integrity counting against them, the forms of deception indulged in by Pasteur and Millikan are likely to continue providing fodder for inquisitive historians.

This is not to say that science need become less rigorous. Nor, as I have already noted, am I proposing the exclusion of either personal ambition or strong emotional commitments to particular ideas. All that is called for is a recognition that leading-edge scientific work justifies being treated rather more indulgently than the vast majority of research that involves consolidating and expanding on existing theoretical frameworks. Seeking to falsify theoretical explanations is essential if science is to be kept free of conceptual detritus. Even so, there is a clear danger that if falsificationism becomes either an end in itself or no more than a defensive strategy deployed by those who fear rival ideas, science, like revolution, will all too frequently consume its own young.

John Snow

Gregor Mendel

Joseph Lister

Charles Darwin

Charles Best

Alexander Fleming

Thomas Huxley

James Young Simpson

P art 1 dealt with the sin of distorting experimental results until they
are consistent with strongly held beliefs. The sins I focus on in
Part 2 are ones perpetrated against history, rather than science. The
general point I'm making is that the way in which each of the above
scientists is generally understood and depicted in textbooks, television
documentaries, and many biographies, is wrong on major points of fact
and interpretation. Heroic caricatures, I seek to show, have displaced
what really happened. Much of the blame for this has to be attributed to
our strong predilection for romanticizing the past. This has led to the re-
shaping of accounts of major discoveries into fireside stories, much richer
in drama than in veracity. But more than a simple desire for entertainment
is at work here. In the next eight chapters I hope to convince you of a far
more fundamental explanation: the strong human tendency to read the
present into the past.

If, following the Whig tradition, we look back in time for the roots of
modern science, there is a serious danger of us wrenching what we find
out of its proper historical context. Words and ideas that meant something only in a given time and place will be cudgelled into becoming
important steps on an upwardly moving staircase of continuous progress.
In these schemas, the direction of advancement is seen as predetermined.
The only uncertainty was on whose brows the laurels would come to rest.
This way of thinking about the past is analogous to the way in which
human evolution used to be conceptualized: with the appearance of

humans as the crowning moment of evolutionary history and the apes, plus a myriad 'lower' organisms, suspended at an inferior level of development.

As the Harvard palaeontologist Stephen Jay Gould has so eloquently described, this approach ignores the fundamental facts that evolution has no fore-ordained goals and no directionality. The only possible definition of an inferior organism is one that fails to keep pace with the changing environment and the demands of competing with its own kind. As humans we may delight in our unparalleled capacity for rational thought, but having a big forebrain is just one of many different strategies for survival. Indeed, should warnings as to the damage we are inflicting on the environment prove as well founded as seems likely, our descendants and those of other species may rue our not having tried something else.

If imputing a predetermined developmental plan makes for bad bioscience, it makes for equally bad history. The inevitable effect of presentism is for the vast majority of human thoughts and experiences to be entirely ignored. Those not ignored are usually gravely misunderstood. In the following chapters I seek to exemplify this by arguing that John Snow, Gregor Mendel, Charles Darwin, Thomas Huxley, and James Young Simpson have all been seriously misconstrued in precisely the way critics of presentism would lead us to expect. The stature of each has been vastly inflated because the parallels between their ideas and those embraced today have been exaggerated by posterity. All highly able men, none enjoyed quite the Olympian detachment that is usually imputed to them today.

Presentism has another negative consequence. Looked at from the privileged viewpoint of the present, many scientific breakthroughs seem so obvious that one is left wondering how our ancestors could ever have been so blind as not to have seen the truth much earlier. This, of course, is to underestimate how profoundly knowledge acquired much later has shaped our own perceptions. Putting aside what we now know is tremendously difficult—as some of the earlier chapters may have demonstrated.

A common result of this is a tendency to accept too readily claims made about our heroes' foresight and prescience. We tend uncritically to embrace the conventional model of the scientific genius as someone who sees something in nature that everyone else was too immersed in alternative ways of thought to notice. Yet, on closer examination we find that some heroes have only been wise after the event. Joseph Lister, Alexander Fleming, and Charles Best typify this. Each was involved in an important development in science and medicine and won great credit. But this was not enough for them. Each also demanded the status of national hero: to paraphrase a remark made concerning Thomas B. Macaulay, 'not content with being a moon' they wanted 'to do a bit in the solar line'. And this is what they achieved. By distorting the historical record, they were successful in persuading posterity to accept their claims to accolades that were largely due to others. The historian Pollio would not have been amused.

MYTH IN THE TIME OF CHOLERA

[John Snow] sat down one afternoon with a map of London, where a recent outbreak [of cholera] had killed more than 500 people in one dreadful 10-day period.

He marked the locations of the homes of those who had died. From the marks on his map, Snow could see that the deaths had all occurred in the so-called Golden Square area.

So Snow went down to Broad Street . . . And, in a gesture that still reverberates among public health scholars today, he removed the handle of the Broad Street pump.

Once the pump was out of commission, the epidemic abated.

Robin Henig, *The People's Health* (1997).

Near Golden Square in central London there is a pub named 'The John Snow' after one of the legends of Victorian science. This is ironic because Snow was a rigid teetotaller and would have set foot on licensed premises only to check their drains. But the reason why the Golden Square pub was so named is clear. John Snow's removal of the Broad Street pump handle has become the stuff of legend.

It seems to be one of those Eureka! moments in which a brilliant scientist spots a pattern never seen before: outbreaks of cholera previously thought to result from contact with airborne filth and foul odours actually cluster around water sources. Next, we have a powerful lead character, who, having stumbled upon the Truth, faces down the prejudices and conservatism of his contemporaries. Rather than see hundreds more needlessly die of cholera, Snow risks humiliation by his very public removal of the handle from the offending pump. Then we have the triumphant vindication sequence in which this beautifully simple action sees Snow

Left: John Snow (1813–58).

widely acclaimed for his genius and courage. As the number of cases of cholera in the vicinity dramatically falls away, it is realized that, at last, the means has been found of controlling this dreadful disease. None could then doubt the scale of Snow's achievement.

The British first experienced large-scale exposure to cholera in India during the late eighteenth century. Moving on from the local population to ravage entire regiments of the India Army, the disease then spread gradually westwards taking the lives of hundreds of thousands of Europeans and, later, Americans. Moreover, only a few years before Snow's researches it had scythed through the French and British armies in the Crimea. Taking many more lives than the Russian guns and sabres, it brought the allies close to defeat. Indubitably, cholera was one of the horrors of the age.

At a more abstract level, from the beginning to the end of the John Snow tale, we are treated to a demonstration of science in its most methodologically pure form. Without preconceptions, Snow plots where the recent cholera deaths are concentrated. His map shows a definite clustering that centres on the water pump in one street. The prevailing theory has it that cholera is an airborne disease, but Snow deduces from his empirical data that it is instead carried by water. A man of action as well as a man of science, he removes the pump handle and the epidemic ceases. QED.

This is an inspiring story; but it is also one in which only the most basic details have any basis in fact. In the summer of 1854 there was indeed a cholera outbreak in and around Golden Square. Its source was correctly identified by John Snow as the water supplied from the Broad Street pump, and the pump handle was removed. In retrospect, Snow was also right in arguing that cholera is usually spread via drinking water polluted with the faecal matter of cholera sufferers. There, however, the genuine details end. The pump handle was removed not by Snow himself but by a local committee set up to deal with the epidemic. This indicates that the significance of the water supply in the context of cholera was already to some extent understood. In addition, by the time the handle was removed from the pump, the water it supplied was almost certainly again safe to drink. The cholera outbreak had precipitately declined in the days before this action was taken and continued to do so afterwards at the same rate. In short, almost every aspect of the Snow story on which emphasis is traditionally placed is mythical.

This begs the obvious question: does this matter? Much as Livy's Horatian myth inspired Ancient Romans, the John Snow story has inspired generations of epidemiologists. It has also enabled Snow's biographers to produce a much more vivid and engaging account of his life. There is no better way of making a subject seem heroic than showing him bucking the system and being proven right in doing so. We have, therefore, to ask whether those who first buy the myth and then sell it on are doing anything more than harmlessly adding colour. In my view they are. Albeit unintentional, the effect of such tactics is to impart a generally misleading idea of how science tends to progress. For this reason, my principal target in this chapter is the rewriting of history in such a way as to give the impression that science is an elite relay race in which the baton of discovery passes from the hand of one giant to the next as the rest of the world gazes on in awe. The Snow myth is the perfect vehicle for this, because whilst John Snow is held up as a pioneering genius, in reality there is hardly an aspect of his thought or action that was genuinely original. Snow may have been, as one historian recently put it, 'a very fine synthesizer' but he was by no stretch of the imagination an innovator.

The causes of cholera

Consider the evidence John Snow accumulated about the Golden Square cholera outbreak. First, on the basis of house-to-house enquiries, he was able to ascertain that Broad Street was the heart of the epidemic. Second, this same information showed him that nearly 90 per cent of those who had died were users of the Broad Street pump. His spot map shows just how compelling this evidence was. Each of the oblong blocks—stacked up like coffins outside individual houses—relates to a separate fatality; the concentrations around the pump are easily apparent. An additional piece of evidence was that many of those who had died some way from Broad Street could also be linked to its pump. For instance, the death of a woman who had lived many years in Broad Street, swore by its pump, and had water brought to her new house in Hampstead from the street

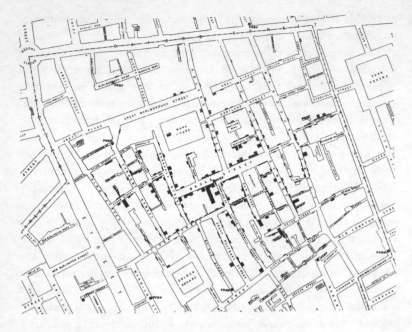

Detail of John Snow's spot map of the Broad Street Pump area.

every day. Hers was the only death from cholera in Hampstead through-out the entire period. So it seems obvious that cholera is a waterborne disease than can best be combated with improvements in sewers, drains, and water supply.

John Snow could cite further evidence in favour of the waterborne hypothesis from an ingenious study he had embarked on in 1853. A cholera outbreak and a privatized water industry gave Snow an unrivalled opportunity. The water market was, as it were, saturated. Competition between different water companies in South London was so intense that in some areas two different companies had laid pipes along the same street. This remarkable situation meant that neighbours might be drawing water from entirely different sections of the Thames. One of these areas struck Snow as particularly auspicious. Single streets were served by both the Southwark & Vauxhall Co., which pumped its water from the filthy waters of the London Thames, and the Lambeth Co., which drew water from further upstream where the water was far less polluted with waste and effluence.

Snow realized that if the users of one of these companies (presumably the Southwark & Vauxhall) were found to be disproportionately subject to cholera, the waterborne thesis would secure a notable victory over the miasma theory, the airborne alternative. He therefore made door-to-door enquiries in the neighbourhood to establish which water company supplied each house and how this correlated with cholera deaths. From a scientific point of view, the results were gratifying. Far more of the Southwark & Vauxhall Co. customers had succumbed to the disease than those supplied from the Lambeth Co.'s much-cleaner water source.

To modern ears, this all sounds conclusive. The facts seem to be speaking at considerable volume for themselves. But if so, why did the Committee on Scientific Inquiries of the General Board of Health adamantly disavow Snow's pump theory? Why, when Snow sent a paper on his studies to the French Academy of Sciences was it ignored and never published? Why, again, was the premier medical journal, *The Lancet*, consistently dismissive of his ideas? Knowing that Snow was correct we have a tendency to dismiss these examples of hostility as ignoble expressions of ignorance, over-weaning conservatism, and vested interest. And, to be sure, some local authorities and sewer authorities were keen to avoid the financial expense of implicating public drains in the genesis of cholera epidemics. But medical observers with their professional reputations at stake were equally insistent on the fatuity of Snow's explanation.

One of these doctors was Edmund Parkes. To understand his critique we must bear in mind that many doctors and laymen during the 1850s quite plausibly thought disease to be caused by 'poisonous miasmas'—usually associated with decaying matter that entered the victim's bloodstream causing inflammation and, on occasion, death. Quite understandably, most struggled to understand how a microscopically tiny particle of matter, or microbe, could cause serious difficulties in an organism as large as a human. Aside from being among the most cited and respected authorities on cholera in Victorian Britain—his *Manual of Hygiene* was the standard text for hygiene specialists and engineers—Parkes was a committed miasmatist. For him, Snow's spot maps showed quite clearly that the incidence of the disease radiated out from an epicentre. As he explained:

> On examining the map given by Dr. Snow, it would clearly appear
> that the centre of the outburst was a spot in Broad-street, close to
> which is the accused pump; and that cases were scattered all round
> this nearly in a circle, becoming less numerous as the exterior of the
> circle is approached. This certainly looks more like the effect of an
> atmospheric cause than any other.

Here we see that the identical spot map was perfectly capable of
supporting the opposing point of view held by a highly intelligent and
distinguished hygienist. Committed to the miasma theory, Parkes was in
no way discomfited by Snow's evidence. Near the pump, he insisted,
there must have been a mass of putrefying matter from which the disease
spread. Parkes was even able to present an imaginative argument against
Snow's version of events:

> If it [i.e. the cholera] were owing to the water, why should not the
> cholera have prevailed equally everywhere where the water was
> drunk? Dr. Snow anticipates this by supposing that those nearest the
> pump made most use of it; but persons who lived at a greater dis-
> tance, though they came farther for the water, would still take as
> much of it. . . . There are, indeed, so many pumps in this district, that
> wherever the outbreak had taken place, it would most probably have
> had one pump or another in its vicinity.

Parkes's testimony shows that when first put forward, Snow's evidence was nowhere near as decisive as is commonly thought. Had we
the capacity to blank out our knowledge of 'what happened next' then
we, too, might struggle to decide which of the theories better fitted the
available facts. Was it the pump, or a nearby pile of putrescence? In mid-
nineteenth-century London there would be no want of the latter and, as
Dr Parkes ingeniously pointed out, a polluted air supply should be
expected to give much stronger clusters than water carried away in con-
tainers.

There is another respect in which John Snow's evidence was insufficient to support the theoretical structure he built on it. Snow did manage
to convince many that the Golden Square outbreak had been caused by
contaminated water. It transpired that one of the earliest victims had been
an infant living at 40 Broad Street. Immediately before its death, the
mother had been in the habit of washing out its nappies in water later

emptied into a cesspool that leaked into the pump reservoir. But converts were not necessarily persuaded that all such outbreaks were waterborne. There was always the possibility that other cholera epidemics were caused in quite different ways. Snow was faced with the logical objection that one cannot generalize from one outbreak to all others without first completing a very extensive epidemiological enquiry. Indeed, proper scientific enquiry demands nothing less. But convinced he was right, Snow insisted on trying to jump the gun. Had his self-assurance been misplaced, he would now be forgotten or reviled; because he was right, he is now revered and his critics reviled. Yet not unreasonably, those critics felt that they were doing no more than their duty in chastising an over-hasty colleague.

In June 1855, *The Lancet* went on the offensive. The journal's founder, the belligerent and radical Thomas Wakley, had been partial to the germ theory that would ultimately vindicate Snow. The leading opinion-formers of Snow's day were passionately opposed. In a stinging rebuke, Snow's logic came under particularly severe scrutiny. The editorial contains an intensity of sarcasm and invective that has now vanished from academic journals:

> **Dr Snow** is satisfied that every case of cholera . . . depends upon a previous case of cholera, and is caused by swallowing the excrementitious matter voided by cholera patients. Very good! But if we admit this, how does it follow that the gases from decomposing animal matter [i.e. miasma theory] are innocuous? We cannot tell. But **Dr Snow** claims to have discovered that the law of propagation of cholera is the drinking of sewage water. His theory, of course, displaces all other theories. . . Therefore, says **Dr Snow**, gases from animal and vegetable decompositions are innocuous! If this logic does not satisfy reason, it satisfies a theory; and we all know that theory is often more despotic than reason. The fact is, that the well whence **Dr Snow** draws all sanitary truth is the main sewer. His *specus*, or den, is a drain. In riding his hobby very hard, he has fallen down through a gully-hole and has never since been able to get out again. And to **Dr Snow** an impossible one: so there we leave him.

This passage has to be evaluated without regard to the eventual vindication of Snow. What is being said of Snow is both rude and cruel; it is also essentially correct. That a water supply contaminated with cholera-infected excreta can pass on the disease does not mean that the health consequences of airborne putrescence can forthwith be ignored. Nor in aggressively pressing this point, were critics solely resting their case on the principles of detachment and caution thought to underpin good science. Miasma theory was very easy to reconcile with Snow's pump hypothesis. For most miasmatists, poisonous odours could readily be carried by water. In fact, it was felt that miasmas were just as likely to arise from water as from mounds of rotting matter. Thus, in May 1850 Dr A. C. MacLaren voiced the common presumption that cholera miasmas were capable of both 'travelling on the winds' and 'shooting along in streams'. Snow's evident difficulty has been summed up by one team of historians and scientists from Michigan State University. In *The Lancet* of 2000 they wrote:

> Some sanitary reformers did find Snow's statistics . . . compelling, but used them selectively to buttress a miasmatist worldview. Dirty water, after all, was just a subcategory of filth, and everyone knew that filth causes disease.

This point serves to underline the inconclusive nature of Snow's data. We can also see that the miasma theory he thought he had consigned to oblivion, was really much more versatile and much more cogent than he admitted. The de luxe miasma theory to which most of Snow's medical peers subscribed was actually perfectly consistent with his evidence and could account for any of his objections. In short, when Snow finally published his data in 1855 it lacked the solidity to overwhelm his critics.

So if John Snow's evidence was inconclusive, what drove him to take an extreme position? Why did he mentally eliminate all forms of contagion except those involving the inadvertent swallowing of faecal matter? The answer does not lie exclusively in the results of his investigations. As we have seen, others who looked at his results remained unconvinced. To fully understand Snow and his critics we need to know the source of Snow's ideas on cholera and, in particular, what he and his contemporaries believed epidemic diseases to be. Investigating these points reveals that Snow was much more a man of his time than is acknowledged by his sizeable modern following.

Cholera and contagion

During the 1850s, medical debates about the cause of disease centred on two main themes: how a disease was contracted and how it affected different people. With respect to contracting disease, the two chief positions were described as 'localism' and 'contagionism'. Both camps were miasmatists in the sense that they saw epidemics as caused by insalubrious local environments producing poisonous miasmas that were inhaled or imbibed by their victims. The 'localists', however, believed that only direct contact with the poisonous, morbific matter could induce illness. In conscious opposition to them, the advocates of 'contagionism' argued that infected individuals could leave the area of original infection and pass the disease to others in the form of morbid matter exhaled or otherwise emanated. Predictably, these contagionists were also enthusiastic proponents of quarantine measures. By the 1850s, after a shaky start, and thanks to several detailed studies of epidemics, they were in a very strong position.

When John Snow came to write his famous book *On the Mode of Communication of Cholera*, published in 1855, he was able to cite dozens of other doctor's reports on the contagious nature of cholera. One of his most striking stories was provided by a Dr Simpson of York, author of *Observations on Asiatic Cholera*. Simpson described how, soon after Christmas 1832, an agricultural labourer called John Barnes, living outside York, died suddenly of a disease that was quickly diagnosed as cholera. This diagnosis presented what seemed to be an inscrutable mystery because no possible source of miasmatic infection could be identified. In the following days, several more local people fell victim to the disease but the doctors were still no closer to identifying the original cause. Then, all of a sudden, the enigma was cleared up. The son of the deceased man arrived in the village and explained that his aunt had recently died of cholera in Leeds. She had no children of her own so her clothes were immediately sent to John Barnes to help him clothe his family. The package of clothes, arriving on Christmas day, had been opened and the garments worn unwashed. This, the doctors realized, was the only possible explanation for the cholera deaths. Tragically, within a few weeks a well-intentioned bequest had killed John Barnes as well as his wife's parents and youngest sister. This story, moreover, was far from exceptional. As one doctor summarized in May 1850, 'Unequivocal instances of contagious communication abound'.

Such examples are important because there is a tendency among those seeking to stress the obduracy of miasma theorists to imply that they were all localists. This misapprehension leads on to the equally false assumption that the whole idea of contagion was unpopular during the mid-nineteenth century. On the contrary. When Snow attacked miasma theory in general on the grounds that only person-to-person contagion could explain what he had observed, he was using an argument that was happily invoked by many miasmatists as well. In sum, John Snow was simply laying an unusual amount of emphasis on just one of the many modes of transmission identified by standard contagionist theory. It wasn't that they were rejecting his ideas out of hand. Instead, without sufficient evidence, he was rejecting out of hand all but a sub-set of theirs.

The germs of the germ theory

Nevertheless, there was more to Snow's theory of cholera than contagionism. After 1849 he also claimed that the cause of cholera was 'particulate', had the power of multiplication, and involved 'veritable animals, or even animalcules'. One of the best accounts of Snow's views is provided by his contemporary, the eminent public-health reformer John Simon:

> This doctrine is, that cholera propagates itself by a 'morbid matter' which, passing from one patient in his evacuations, is accidentally swallowed by other persons as a pollution of food or water; that an increase of the swallowed germ of the disease takes place in the interior of the stomach and bowels, giving rise to the essential actions of cholera, as at first a local derangement; and that 'the morbid matter of cholera having the property of reproducing its own kind must necessarily have some sort of structure, most likely that of a cell'.

Even though the cholera *Vibrio* was not isolated until 1883 by the German Robert Koch, Snow was moving rapidly in the direction of the modern germ theory of disease. At the same time, he was jettisoning another ancient belief. Many of Snow's contemporaries were convinced that there was no such thing as a specific disease-causing agent that always produced an identical ailment. The same miasma, they claimed, might cause one person to fall victim to influenza, another to chronic diarrhoea, and yet another to cholera, according to their physical condition, state of mind, and, especially, the nature of their inborn 'constitution'. This was the second major theme mentioned above. The theory of the individual constitution dispensed with the need to identify the cause of particular diseases by arguing that different ailments are no more than each person's individual response to noxious agents. Conversely, for Snow, a specific agent caused cholera and it always did its damage in the gastrointestinal organs.

Yet, once more Snow's thinking was far from new. Even if it didn't represent the majority opinion, the microbial theory of cholera had been well rehearsed in the medical literature of his day. The German-born doctor Jacob Henle had adumbrated modern germ theory in a paper of 1840, in which specific 'parasitic organisms' were implicated in the cause of 'specific' infectious diseases. Within a few years, the role of several multicellular parasites in producing human and animal disease had been widely accepted. And in 1842, the Scot John Goodsir created something of a sensation by showing the existence of highly distinctive bacteria in the vomit of some of his patients. Furthermore, by the 1850s the growing understanding of smallpox lent considerable credibility to the notion that disease involves organic matter that can multiply itself and produce predictable symptoms in all its sufferers. Few denied that smallpox lesions only give rise to more cases of smallpox when transferred from arm to arm. Nor did many dispute that in order for vaccines to work, the infectious material had to be kept fresh. This all served to undermine the theory of constitutions and provided a fertile ground for the emergence of germ theory. By 1848, Snow was already able to draw on a widespread willingness to extrapolate from smallpox to the rest of the disease taxonomy.

The idea that cholera was a gastrointestinal illness contracted via contaminated water also pre-dated Snow's taking up the cudgels on its behalf. During the 1830s, the Frenchman François Leuret compiled substantial evidence in its favour. Then, in September 1849, well before Snow's Golden Square enquiry, another doctor-hygienist, William Budd, wrote in *The Times* that cholera is caused by 'a distinct species of fungus which, being swallowed, becomes infinitely multiplied in the intestinal canal, and the action thus excited causes the flux of the cholera . . . water is the principal means of the dissemination of the disease'. Budd's description of cholera as a disease spread by the faecal–oral route coincides almost precisely with that advanced by Snow himself. But it cannot be argued that it was Snow who influenced Budd. A Devonian by birth, Budd practised in Bristol and remains a local hero on the basis of the practical hygiene measures he championed during the 1850s. With such schemes as providing depots where Bristolians could obtain free disinfectant, it is said that he reduced the death toll during the next cholera epidemic by over 90 per cent.

Budd's (and Snow's) thinking also overlapped with that of the British sanitary engineer Henry Cooper. In 1850, Cooper published a detailed study of a cholera epidemic in Hull. Although he was not inclined to associate cholera with only one mode of transmission, he emphasized the relationship between the intensity of outbreak and the 'efficacy of drainage'. The 1849 epidemic, he explained, had convinced the local Board of Health to institute a large-scale programme of sanitary improvement. Similarly, in 1850, after a cholera outbreak in Salford, the statistician A. C. MacLaren delivered a paper in which he, too, argued that cholera may be spread by infected water. MacLaren also shared the conviction of Budd and Snow that cholera is a specific gastrointestinal infection. The 'gastro-choleric irritation is specifically cholera' he stated before a prestigious scientific gathering in London.

Likewise, the drama of taking away the pump handle cannot be described as without precedent. The association between contaminated well-water and cholera had already been made in America, where pump handles had been removed during several cholera epidemics. Closer to home, at the height of the 1850 outbreak in Salford, investigators are known to have impounded local pump handles. As in Broad Street 4 years later, this action came to be associated with the outbreak's dying out. Given the natural life cycle of epidemics, however, such linkages may be spurious. Certainly, as has been pointed out, the Broad Street death rate was already in steep decline before the pump handle was taken away.

So, in seeking to understand the origins of John Snow's theory of cholera transmission, we can conclude that his opinions were exceptional only in the vehemence with which they were held. Similar beliefs were expressed by other leading hygienists, and Snow was certainly influenced by the ideas of earlier and contemporary medical writers. The claims that cholera is generally waterborne and involves infectious particles that can only produce cholera symptoms were common to a significant and vocal minority of doctors during the early 1850s. Therefore although Snow deserves much of the fame bestowed on him—for the reasons I discuss below—he cannot accurately be characterized as a pioneering genius.

The myth of scientific detachment

One serious criticism that might be made of Snow is that he lacked scientific detachment. Far from carrying out his enquiries free of preconceptions, his own records show that he was utterly convinced that cholera spread contagiously, via the drinking of water infected with faecal matter before he embarked on them. Just like his medical opponents, Snow was looking at the data from within his own framework of theories, ideas, and axioms. This is why when he looked at his spot map he could arrive at a conclusion that was vigorously disputed by many of his contemporary hygienists even when the map was made available to them.

As I explained in Part 1, however, this commitment to prior theory should earn Snow no retrospective censure. Over the past few decades there has been a general realization that scientists rarely look at nature without having a theory to guide them. This is mostly because the external world is bewilderingly complex. A mind incapable of imposing preconceived ideas onto what it observes will soon implode under the pressure of sensory overload. In the jargon of the philosophy of science, our perception of natural phenomena is 'theory-laden', and cannot be otherwise. The only alternative to the despotism of preconceived ideas is, in nearly all cases, an anarchic haze of disconnected thoughts and impressions. Theories also play an essential part in making linkages between seemingly disparate observations. Without a theoretical framework within which to make sense of them, many crucial facts seem isolated and unimportant. For instance, *Archaeopteryx*, the fossilized 'missing link' between reptiles and birds, was discovered just 2 years after the publication of Darwin's *Origin of Species*. The importance of the find was recognized only because naturalists were suddenly on the lookout for evidence of evolution. In a similar fashion, John Snow's great strength as a scientist was his prior commitment to a well-defined set of ideas. It was these that drove his superbly contrived investigations.

Was John Snow the father of epidemiology?

What we can now be reasonably certain of is that Snow's pump theory did not emerge in pure form from his famous spot maps. Contrary to the textbook view, Snow had identified the pump water as the arch-culprit more than 3 months before drawing up his first such map. Indeed, the diagnostic value of epidemiological maps grew on him only slowly and for the most part he used them just for purposes of illustration. The first edition of his *On the Mode and Communication of Cholera* contained not a single map and only one table. It would seem, then, that his reputation as the 'father of epidemiology' may be amongst the most inaccurate aspects of the John Snow legend. We now know that by the time he started to think of contagious disease in topographical terms (that is, linking the distribution of deaths to aspects of the local environment), he had a lot of catching up to do. Just how much is indicated by the genuinely pioneering work in the late eighteenth and early nineteenth centuries of such men as Valentine Seaman, Benoiston de Chateauneuf, Michel Chevalier, William Farr, George Busk, Thomas Shapter, Henry Cooper, and George Mendenhall. If we set about the very questionable task of establishing the paternity of the discipline, many—if not all—of these men will be shown to have a better claim than Snow. The only advantage he enjoys is a better legend.

In France, with a powerful state intent on the efficient mobilization of its resources, both Benoiston de Chateauneuf and Michel Chevalier undertook detailed epidemiological surveys. Chevalier's 1830s studies showed especially striking correlations between disease incidence and the poor neighbourhoods of Napoleonic Paris. In Britain, William Farr—one of the nineteenth-century's greatest practical statisticians—mounted some of the most elaborate and complex epidemiological surveys from the 1840s onwards. Formidable in both appearance and intellect, Farr used his access to the nation's vital statistics to defend the miasma theory.

In 1852, he published a paper showing a very high inverse correlation between altitude and the probability of contracting cholera. He plausibly reasoned that low-lying areas contain more organic matter capable of releasing noxious miasmas. Although in later years he would recant and become an enthusiastic advocate of the germ theory, a properly contextualized judgement allows us to admire the cogency of his original argument.

Nor was John Snow the first to use the specific tool of the spot map. In 1798, an American doctor named Valentine Seaman had used two spot maps of New York to study the deaths from an epidemic of yellow fever. For Seaman, as for most of his successors, spot maps provided a useful means of supporting miasmatist theories of contagion. In 1849, the sanitary engineer Henry Cooper also used spot maps to great effect in his study of Hull's 1849 cholera outbreak. As he explained in a paper read to the Statistical Society of London, 'the dots or dark marks are placed wherever cases of cholera occurred . . . there dots will, by their aggregation, show strikingly the spots of greatest mortality'. Soon after the Golden Square epidemic began, this same Henry Cooper was called in to investigate the rumour that the construction of local sewers had disturbed an ancient burial ground for those who had died in the plague year of 1665. Many locals were convinced that this had released offensive gases into the atmosphere through the gully holes that stretched up from the underground sewer network. Cooper at once drew up a spot map—the very first to refer to Golden Square and Broad Street—from which he could show that there were fewer than average deaths from cholera in the vicinity of the plague pit. He also revealed that the sewers laying adjacent to the pit actually ran north into Regent Street, where there was no sign of cholera. Unlike Snow, then, Cooper used his spot map as an analytical tool.

Snow the synthesizer

So far the main thrust of the argument has been that the canonization of individuals like Snow makes for bad history and a distorted understanding of the way in which science usually works. There is also another problem. Overstating the importance of a long-dead scientist carries with it the risk that when the truth is finally revealed there can be what the military call 'collateral damage' to the scientific discipline.

Take, for example, the statistician and 'measurer of minds' Cyril Burt. For most of his professional life he was treated as the doyen of IQ research. The data he had obtained over a lifetime's research into IQ heritability was by far the most influential and compelling in the field. Then, shortly after his death, he was exposed as an academic fraud. Not only was he accused of having concocted data, he seemed to have invented the existence of the technicians who were supposed to have collected it. Burt's fall from grace was rapid and unimpeded. And although the case against him is unproven, the allegations enabled the numerous critics of hereditarian research to broaden the attack to include the entire discipline with which Burt was so strongly associated. Even Hans Eysenck, an avowed hereditarian, acknowledged in his autobiography that the publicity surrounding the affair led many people to reject 'any theories concerning the inheritance of intelligence'. The message seems to be clear. No one individual—alive or dead—should be elevated to such a degree that their reputation becomes indistinguishable from that of their discipline.

Were Burt not warning enough, the current question marks over the anthropologist Margaret Mead's *Coming of Age in Samoa* seem to be another case in point. Some anthropologists have been acutely embarrassed that some aspects of Mead's 1920s research might have been fabricated. Surely, then, a fully mature and justifiably self-confident discipline has no need of an irreproachable father/mother figure.

Nor is it reasonable to suppose that there could ever be such a being. One of the reasons individuals such as Burt and Mead fell so far is that most reappraisals start from an exaggerated idea of how disinterested the archetypal scientist is and how much great scientific work one individual is actually capable of doing. In other words, discredited scientists fall from pedestals of our own construction, which are themselves balanced atop largely erroneous beliefs about the role of genius in the history of science.

If, instead, we accept that the image of the lone genius is nearly always a wild exaggeration, we can come to see individuals like John Snow as genuinely great even though they themselves made only small steps forward. Once we see that great strides rarely occur, we acquire a more realistic concept of what 'greatness' actually involves. Between 1848 and 1855, Snow adopted a specific theory of cholera pathology and he investigated the incidence of cholera using epidemiological tools. Neither theory nor methodology was new. But now that we have scaled down our expectations, this matters far less. Snow's analysis of the Golden Square epidemic and, to an even greater degree, his investigation of the South London water supply were stunning examples of the synthesis of theory and methodology. He may not have conclusively proved that the faecal–oral route was a basic feature of cholera transmission. Yet, one could hardly invent a clearer demonstration of the fact that cholera can be waterborne than that provided by the brilliant use he made of the inter-mingling of Lambeth Co. and Southwark & Vauxhall Co. water con-sumers. It is hard to say how influential either of his studies was. Nevertheless, judged according to the scientific standards of both his time and ours, Snow was an exceptionally talented and ingenious researcher.

'THE PRIEST WHO HELD THE KEY'

Gregor Mendel and the ratios of fact and fiction

In 1900, three biologists independently rediscovered Mendel's laws, according to which the characteristics of organisms are determined by hereditary units, each kind being present once in a gamete, sperm or egg, and hence twice in the fertilized egg. In effect, it was the atomic theory of heredity.

> John Maynard Smith, *New York Review of Books*
> (21 December 2000).

Mendel's genius was not [the] flamboyant, touched-by-an-angel kind. He toiled, almost obsessively, at what he did. But still he had that extra one percent, that inspiration that helped him to see his results in a way that was just slightly askew. This flash of insight—even if it follows long, dull stretches of routine labour—is what made Mendel great. It allowed him to perform a feat of genius: to propose laws of inheritance that ultimately became the underpinning of the science of genetics.

> Robin Henig, *The Monk in the Garden* (1999).

'It is not often possible to pinpoint the origin of a whole new branch of science accurately in time and place . . . But genetics is an exception, for it owes its origin to one man, Gregor Mendel, who expounded its basic principles at Brno on 8 February and 8 March 1865.' This proud declaration was made by the British evolutionist Sir Gavin de Beer as the cream of the genetics community celebrated the birth of their discipline. It was April 1965 and a day of high emotion and tremendous pride. A hundred years had passed since a comparatively obscure Moravian monk had published a paper that would revolutionize the study of heredity. De Beer enthused over the way in which, in a small monastic herb garden,

Left: Gregor Mendel (1823–84).

Mendel had personally discovered the basic principles of heredity that had eluded mankind for thousands of years.

Yet, de Beer lamented, Mendel's own life contained far more tragedy than glory. His brilliant paper was almost entirely neglected by his contemporaries and only after Mendel's death was its massive potential realized. It then caused an intellectual explosion so powerful that it propelled the fledgling science of heredity into hitherto unimaginable pre-eminence. De Beer explained that before the rediscovery of Mendel's ideas, hereditarian thought had been little more than a miscellany of bogus laws, spurious observations, folklore, and haphazard rules of thumb. Once Mendel's ideas had been clearly understood, a fully fledged science came into being. And the profound implications of his ideas for the whole human race became ever more obvious with each year that passed. According to de Beer's reading, if one of the measures of genius is just how far 'ahead of his or her time' a particular thinker is, then Mendel has to be credited with at least 40 years. His epoch-making paper, first presented in 1865, was rediscovered and embraced only in the early 1900s. By this measure alone, he was a scientific revolutionary of the highest order.

It was not only geneticists that felt the need to pay homage in 1965. Mendelian ideas had also revitalized evolutionary theory by providing what Charles Darwin so desperately sought: a plausible theory of heredity that made biological sense of natural selection. Here, de Beer explained, the history of biology revealed its cruellest twist. Mendel's paper was published in a relatively obscure Austrian journal and, as a result, never came to Darwin's attention. Had he read it, the genetic underpinnings now seen as crucial to evolutionary biology could have been acquired almost half a century earlier. In that anniversary year, the American historian Loren Eiseley made much the same point. Eiseley dubbed Mendel 'the Priest who held the key to evolution' and said that because Darwin and Mendel tragically passed each other like ships in the night, biology had to endure a long and unnecessary impasse during which faith in Darwinism itself largely collapsed. 'No man who loves knowledge would want an episode like this to happen twice', Eiseley gravely concluded.

De Beer's audience and Eiseley's readers willingly accepted these ringing endorsements of Mendel's status as the founding father of genetics.

Yet, it is now apparent that both men were liberally applying the varnish to a myth that had been developing around Mendel's name since the beginning of the twentieth century. Following the work of several modern historians, I argue in this chapter that the nature of Mendel's contribution to science has been misunderstood for more than a century. In life, this monk–cum–scientist was a tenacious researcher ultimately frustrated in his attempts to breed new, stable plant varieties from hybrids; in death, he was granted one of the highest benefices in the hallowed realm of scientific thought.

As I draw on recent historical scholarship to challenge this extraordinary rise to glory, the pertinence of an observation made by the British statistician and Darwinist, Sir Ronald A. Fisher will become very clear: 'Each generation [has] found in Mendel's paper only what it expected to find', he wrote in a 1936 edition of *Annals of Science*; 'each generation, therefore, [has] ignored what did not confirm its own expectations.' In other words, the prevailing accounts of Mendel's career betray all the classic weaknesses of presentism. It would be hard to imagine a finer demonstration of the cumulative effects of succeeding generations projecting the present back onto the past than is to be found in our current image of Gregor Mendel.

Classical Mendelian genetics

To make clear just how much myth has coalesced around Mendel's name, I will first outline what he is famous for. His towering reputation rests on three things: the two scientific laws that bear his name and the discovery of what are called the Mendelian ratios.

The first law is that of segregation. Understanding what it means requires familiarity with a few technical terms and some basic facts of reproductive biology. The key terms are gametes, chromosomes, and alleles. 'Gametes' encompasses both sperm and eggs—the cells that are central to sexual reproduction. 'Chromosomes' are the strands of DNA on which genes are located. 'Alleles' are gene variants any of which can occupy the same position on a set of chromosomes; for example, a gene coding for brown eyes is allelic to one coding for blue.

The way these different entities interrelate will become clear if we consider the two types of cell present in the human genome. Most are 'somatic' (or body) cells, the biological building blocks of which we are constructed. These contain two complete sets of the 23 chromosomes on which the whole human recipe is carried—a total, therefore, of 46 chromosomes: one set is derived from one parent, its pair from the other. The individual's genes are arrayed on these complementary chromosomes, each gene being paired with its equivalent on the other set of chromosomes. In many instances, the two genes comprising the pair are virtually identical; in others, the pair comprises two different alleles.

The second type of cell are gametes, the sexual reproduction cells already referred to. Although derived from somatic cells, gametes differ in that they contain only one set of chromosomes ($n = 23$) and hence only one set of genes. Mendel's Law of Segregation rests upon the way in which gametes are formed from normal body cells. The gene pairs that lie side by side on matched pairs of chromosomes in a somatic cell are broken up as the two sets of chromosomes are first copied, then shuffled together, and finally separated into four new sets, each of which forms a gamete. In the final, fertilization, stage of this process a male's sperm and a female's egg couple their individual sets of 23 chromosomes to produce another generation of somatic cells with the requisite double set of chromosomes. Humans wait (on average) for the best part of two decades and then the whole show usually passes through another cycle.

This may seem like jargon-ridden technicality, only of interest to life scientists. In reality, though, it is a process central to human existence and individuality. It also tells us two important things about the nature of genes. First, that they are distinct entities; and, second, that they are in no way altered by being held in such close proximity to their own kind. As will be seen, an allele for brown eyes will suppress the expression of one coding for blue eyes if it is paired with it. Nonetheless, once reproduction has copied two 'blue-eye' genes into an embryo to the exclusion of their 'brown-eye' rival, blue eyes re-emerge in all their glory. As the British zoologist Richard Dawkins pointed out in *The Selfish Gene* (1976), compared with our brief lives, 'genes, like diamonds, are forever'. What Mendel's Law of Segregation stipulates is that the two members of each parental gene pair going their separate ways is an essential element of sexual reproduction. No matter how much reshuffling takes place, no erstwhile members of any given gene pair will ever find themselves in the same gamete.

Mendel's second law is closely related to the first. Called the Law of Independent Assortment it states that the physiological independence of genes is such that each gamete will contain a random mixture of chromosomes derived from the carrier's paternal and maternal genomes. There will be only one of each sort of chromosome (and therefore allele), but whether that chromosome was present in the individual's father or mother is purely a matter of chance. We can take an example from Mendel's own work. Domesticated pea plants can produce smooth seeds or wrinkly ones. The seeds may also be either yellow or green. Assuming two sets of rival alleles to be at work here, Mendel's second law tells us that whether the gene for wrinkly seeds or the one for smooth seeds makes it into any particular gamete, this cannot influence the outcome of the 'rivalry' between the two genes coding for colour. For example, the combination of smooth and green is just as likely as smooth and yellow. In short, genes are fully independent travellers.

We now know, however, that this law comes with a caveat. As I have pointed out, the reshuffling that goes on before gametes are formed is of chromosomes, not genes. As a result, genes on the same chromosome are quite likely to be prolonged travelling companions. If, say, genes determining eye colour and straightness of hair were located on the same chromosome, brown eyes might prove a very strong predictor of curly hair. But even this is subject to qualification. Natural selection so favours variation that the means whereby pairs of chromosomes can mix and match between themselves have evolved. As a result, we can safely say the Law of Independent Assortment is honoured a great deal more in the observance than the breach.

We now have only Mendelian ratios to deal with. These were made apparent to Mendel because in first breeding between pure strains of pea plant to produce hybrids, he obtained specimens in which some gene pairs comprised easily distinguishable alleles. For example, he bred purple-flowered pea plants with ones having white flowers. Crucially, whereas some paired alleles work co-operatively and produce intermediate colours, colour genes in pea plants work on the principle 'winner takes all': what is called the dominant allele is always expressed at the expense of the recessive form. In pea plants, purple has dominance over white. Already familiar with this, Mendel bred between pairs of the first-generation hybrids. From this he obtained his first ratio: on average, out of every four plants produced, three were purple and one was white. Cross-breeding with these enabled him to refine his 3:1 into 1:2:1. This was because, whereas the white plants always went on to produce solely white plants, only one of the purples consistently produced purple offspring. He therefore knew that his initial 3:1 really comprised one true purple, two hybrids, and one true white.

How much more Mendel learned from this is considered later in the chapter. Here it is worth spelling out why the 1:2:1 is so important. It is a convention of biological notation that a dominant allele is denoted with an italicized uppercase letter, the recessive with an italicized lowercase letter. So if we stick with colour in hybrid pea plants we can represent purple, the dominant colour, with *C* and white, the recessive, with *c*. The gene pair for colour in two hybrid parents can therefore be denoted by *Cc*. Knowing these parental gene pairs enables us to make use of a simple matrix devised by a Cambridge mathematician called Reginald Punnet shortly after the 'rediscovery' of Mendel's work. Punnet Squares are made up of four cells: one parent's gene pair is put along the top and the other's down the side. As shown opposite, this arrangement not only yields all the possible permutations chance could throw up but also the frequencies at which they are likely to occur:

Our four cells give us one *CC*, two *Cc*s, and one *cc*. Because reproduction entails a lot of chance events, this *CC* + *2Cc* + *cc* ratio is not guaranteed to be found in each set of four young. But given the number of plants with which Mendel was dealing, its presenting itself to an observant researcher working with a plant as revelatory as the edible pea was always likely. For the future of genetics, the great challenge lay in reading back from the physical evidence of the three purple-flowered plants and one white produced by the two purple hybrid parents to appreciate that, in somatic cells, units of heredity (genes) must routinely co-exist in pairs. Put another way, the challenge for Mendel was to deduce from the Mendelian ratios what was to him the non-observable fact that the parent plants carried sets of gene pairs rather than a single set of genes, and then

Parental genes	C (dominant gene for purple colour)	c (recessive gene for white colour)
C	CC dominant	Cc hybrid
c	Cc hybrid	cc recessive

A typical Punnet Square.

go on to realize that this might be the standard arrangement, not something peculiar to hybrids.

The summary above is all we need to know about Mendelian genetics for present purposes. Our task is now that of the historian of science: seeking to gauge just how much of this information would come as a complete surprise to Mendel were we able to raise him from the dead.

What was Gregor Mendel doing in 1865?

Having researched this issue in great depth, the British historian Robert Olby has recently raised the question of whether Mendel himself can be said to have been a 'Mendelian'. By this he means that the bulk of the ideas outlined above, and attributed to Mendel in almost every available biology textbook, would have astonished and mystified Mendel himself. This is striking stuff. Could historians and biologists really have had it all wrong for over a century? To check this out by following the trail blazed by Olby, we need first to understand why Mendel started his work with pea plants during the late 1850s. If we do so, it rapidly becomes clear that what he most certainly was not about was discovering the laws of heredity. In fact, the scientific framework within which he was working has no modern equivalent: Mendel devoted most of his scientific life to what eventually proved to be an intellectual dead-end. Like others of its kind, because its sterility has long since been recognized, little or no space is now found for it in standard historical accounts. But to fully understand what Mendel was doing, we need to get to grips with it.

Let's start with the title of Mendel's most famous paper: 'Experiments in plant hybridization'. Note that it was not 'The laws of hereditary transmission' or 'The mechanics of heredity', nor even 'Heredity in *Pisum sativum*'—the edible pea variety on which he was experimenting. The word 'hybridization' crops up repeatedly in Mendel's writings whereas 'heredity' hardly appears at all. At the very least this is highly suggestive. Next we may examine the essay's introduction: what did Mendel say that he was doing? On this he was explicit. Mendel claimed to be presenting the results of 'a detailed experiment', the aim of which was to establish a 'generally applicable law governing the formation and development of hybrids'. It is much the same story at the end of the paper. He makes no claim there to have deciphered the statistical laws of heredity. Instead, he declares that he has shed light on the opinion of a botanist called 'Gärtner' who 'was led to oppose the opinion of those naturalists who dispute the stability of plant species and believe in a continuous evolution of vegetation'. For us, there is only one difficulty: what exactly does all this mean?

A brief excursion into eighteenth- and nineteenth-century botany enables sense to be made of it. During the 1860s, Mendel was actively exploring an issue that once galvanized the botanical community. It was first raised by the famous Swedish taxonomist Carl (or Carolus) Linnaeus, the originator of the still-current system of classifying organisms in terms of species and genus.

During the 1750s, Linnaeus began to doubt that species are as immutably fixed by Creation as religious orthodoxy insisted. His doubts were fed by the wonderfully exotic forms of flora that were being brought back to Europe by explorers. As a taxonomist, whose job it was to pigeon-hole living forms in neat categories, this was little short of a nightmare. The quantity and variety of these new plants and animals soon over-whelmed the existing European classificatory frameworks. And as he struggled to re-impose order, Linnaeus could not help but be impressed by the undeniable evidence of the plenitude of nature. Soon he was thinking the hitherto unthinkable. Had God really made all these diverse species in one brief creative episode or had many of them been formed from an originally much-smaller suite of primordial forms?

Slowly Linnaeus began to favour the second, evolutionary, possi-bility. But the evolutionary mechanism he proposed bore no resemblance to Darwinism. He gave no thought to environmental pressures or to the appearance of random variation. His interest centred on the well-known botanical phenomenon of cross-breeding between different varieties. As this demonstrably led to the emergence of novel plant forms, he speculated that, interbred for enough generations, such hybrids might eventually become entirely new species. Known as 'species multiplication by hybridization' this was an idea that consumed a great deal of scientific effort over the next century. Not least because of its implications for commercial plant breeders, at different times Holland, France, and Prussia offered major prizes for the paper that most effectively addressed the question. Rather than vindicating Linnaeus's idea, however, investigators repeatedly found themselves unable to stabilize hybrid forms. Over and over again, the offspring of hybrids either returned to parental forms or, suffering from lowered fecundity, died out.

Despite this, species multiplication by hybridization seems to have been one of those areas in which scientific hope springs eternal. Throughout most of the nineteenth century there were always some botanists still convinced that they could produce what had become known as 'constant hybrids'—that is, hybrids that had become new species. Indeed, while Mendel was at university in Vienna, the botanist Franz Unger assured him directly that hybridization might well be a source of new species. As we have no reason to believe that Mendel's religious commitments were not real, it is unsurprising that he took up this area of study. Far from being seen as allied to the blind forces of Darwinian evolution, variation by hybridization was thought of as an increasingly necessary adjunct to the Creation story. After all, what could better demonstrate the sublime skills of the Creator than evidence that a judicious, but modest, initial endowment of plant life had been given the in-built capacity to bring forth near infinite variety?

This, then, was the once important botanical tradition to which Mendel's 'Experiments in plant hybridization' were a contribution. Hybrids were Mendel's chief point of interest not as a useful means of investigating the dynamics of hereditary transmission but as a way of vindicating Linnaeus's 100-year-old speculation. Once this is realized, the meaning of previously obscure passages in Mendel's papers become perfectly clear. Mendel was convinced that hybridization permits a 'continuous evolution of vegetation' and the aim of his experiment was to breed hybrids together, generation after generation, to see if they became a new species. This is why he systematically culled any hybrids that had bred with pure-type peas, proved infertile, or did not grow well. His 1865 essay is really a detailed record of his attempts to produce new species. And proving Linnaeus right was of such importance to Mendel that he significantly misrepresented the views of one of his experimental predecessors.

To buttress his contention that hybrids could become species, Mendel claimed that Max Wichura, a world authority on willows, also believed that willow hybrids 'propagate themselves like pure species'. There is a serious difficulty with this. When Robert Olby went back to Wichura's original papers he found that Wichura had repeatedly shown that hybrid willows have a strong predilection for reverting to their ancestral forms. As a result, despite Mendel's crediting him with the opposite view, Wichura remained very doubtful as to the credibility of Linnaeus's hypothesis.

Unfortunately for Mendel, try as he might, his hybrids also showed an unremitting capacity for reversion to the original parental forms. Modern genetics tells us why. He was engaged in an unequal struggle with the effects of dominance and recessiveness on dissimilar gene pairs. As we have seen, this is a process that makes it inevitable that, generation after generation, only half the new plants produced from hybrids will be hybrids. Whereas pure forms fertilized by one of their own kind produce nothing but identical pure forms, the hybrid can never stray far from the 50 per cent rule. As the other half of its output will be true-breeding pure forms, the overall ratio of hybrids to pure forms progressively diminishes. Even the distinctive appearance of hybrids is illusory. Hybrids owe it not to genes peculiar to their hybrid form, but to the interactions of the two alleles separately responsible for the original parental pure forms. In short, Mendel's own work showed conclusively that no hybrid lineage is capable of forming only hybrid offspring.

This was a dismal outcome for a scientist trying to prove that hybrids can produce entirely new species. It is reported that Mendel was by nature taciturn, but some of his frustration with these results seems to have found its way into the account he gives of his work. We can clearly discern this in Mendel's most famous paper: his 1865 'Experiments in plant hybridization'. Closing the essay, Mendel tried to wriggle free of his data. Arguing that his experiments were inconclusive, he lamely conclud-ed that the results were not so clear that they had to be 'unconditionally accepted'. Despite all the evidence to the contrary, at the time of writing he remained committed to the belief in the possibility of 'constant hybrids'. An appreciation of this fact calls into question the standard version of Mendel's famous presentations to the Brno Society for the Study of Natural Science in 1865.

Loren Eiseley, convinced of his subject's unparalleled prescience, described this meeting thus:

> At the end of the blue-eyed priest's eager presentation of his researches, the still existing minutes of the society indicate that there was no discussion. . . . No one had ventured a question, not a single heartbeat had quickened. In the little schoolroom one of the greatest scientific discoveries of the nineteenth century had just been enun-ciated by a professional teacher with an elaborate array of evidence. Not a solitary soul had understood him.

Read in conjunction with Olby's work, what Mendel's papers actually say suggests a very different picture. Given that Mendel had first entered the local monastery over 20 years earlier and had been engaged in his work with pea plants for about a decade, it seems likely that many of those listening to him knew what he was about and what he had been hoping to achieve. The huge presentist superstructure removed, we can now see that, in 1865, Mendel was actually reporting unequivocal failure. He had thrown up an interesting statistical pattern that he could not fully explain, but even his practical hopes of finding a way of stabilizing new plant types for the benefit of local farmers had got nowhere. If we assume that many among his audience were aware of this, we could reread their silence as reflecting not incomprehension but sympathetic understanding.

Ratios of dominants and recessives

Now that we understand what Gregor Mendel was doing and the intel-lectual framework within which he was working, we can start to gauge the extent to which he directly contributed to the development of modern Mendelian genetics. From the outset it is clear that one component of the modern discipline is contained within his 1865 and 1866 essays. Mendel did have a clear sense that some characteristics are dominant and some are recessive. Early in the 1865 essay he explained that:

> Characters which are transmitted entire, or almost unchanged in the hybridization, and therefore in themselves constitute the characters of the hybrid, are termed the *dominant*, and those which become latent in the process *recessive*.

Indeed, I made the point earlier that Mendel's entire experiment revolved around an appreciation of this. He bred 10 000 plants of several edible pea varieties (*Pisum sativum*) into pure strains. Then he mated these different kinds of pure-type pea together to form hybrids. His focus of attention was any characteristic that differed markedly between two pure strains. I have already referred to purple versus white flower colour and smooth versus wrinkled seeds. Tall plants versus short plants is another example. Mendel intentionally selected couplets in which he knew that one trait was dominant and one recessive and that the hybrid form somehow contained both.

Of itself, however, this does not mark out Mendel as an original thinker. We must not unthinkingly assume that to be aware of the phenomena of dominance was to enter new scientific territory. When Mendel started his experiments in the mid 1850s, the idea that some characteristics were dominant over others was common knowledge amongst botanists. In the early years of the century the British botanists Thomas Andrew Knight, John Goss, and Alexander Seton had written extensively on the subject; it was a staple of the papers of Europe's leading theorists; and Darwin routinely referred to the 'pre-potency' of certain traits in his own plant and animal breeding experiments.

As I have already indicated, the big leap that was waiting to be made was the realization that the out-workings of the relationship between dominants and recessives reflect the existence of gene pairs. Further, that far from being solely of relevance to species with dominant and recessive characteristics, gene pairs are the standard arrangement. Our ability, therefore, to credit Mendel with having the necessary insight to have formulated the Law of Segregation that carries his name must turn on the question: did he complete these additional steps? After all, if there were not gene pairs, there would be nothing to segregate and nothing to legislate on. This is a question we are now in a position to take up.

Segregation, characters, and elements

As we have seen, there can be no doubt that Gregor Mendel had a good, if unexceptional, grasp on the external effects of dominance. He had also discovered that, with pea plants at least, hybrids produce a 1:2:1 ratio for certain characteristics expressed in their offspring. So next we have to ask to what extent did he understand the role of gene pairs in the genetics of the edible pea? In trying to answer this, the first point that has to be made is that, except in rare instances, Mendel did not talk about the mechanisms of heredity at all. This is a bold claim that needs to be well backed up. We begin with one of the key sentences in Mendel's 1865 paper. Here, he describes what he thinks happens when two different pure types are mated together, producing a hybrid form. Note very carefully the language he uses:

> If **A** be taken as denoting one of the two constant characters, for instance the dominant, **a** the recessive, and **Aa** the hybrid form in which both are conjoined, the expression: $A + 2Aa + a$ shows the terms in the series for the progeny of the hybrids of two differentiating characters.

Something immediately jumps out about his phraseology: where we would use the word 'genes', Mendel uses the German for 'characters'. Of course, one would not expect him to use the word 'genes' (it was not coined until 1903). Yet, Mendel had his own word—'elements'—for the particles of hereditary material. And in his entire 1865 essay he used the word only 10 times. In contrast, he used the term 'characters' on 182 separate occasions.

This would be unimportant were it not for the fact that Mendel used 'characters' only when referring to physical characteristics, and never when discussing the contents of the reproductive cells. Conversely he used the word 'elements' only when talking about the discrete particles of hereditary matter that passed from one generation to the next. As in the above passage, when Mendel paired up dominant and recessive traits, as in '**Aa**', he was always referring to 'characters' and never to 'elements'. Of course, Mendel realized that these 'characters' were inherited. But the key point is that he kept his discussion at the level of the physical characteristics themselves: paired characters in Mendel's terms did not equate to paired genes. Why this should have been the case is obvious: it's what he could actually see. Observing the offspring of hybrids told Mendel that the parent plants contain the hereditary potential for two different forms of traits, such as size and colour. Yet simply looking at hybrids and their progeny didn't open his eyes to what was going on within the reproductive cells themselves.

It is just as significant that Mendel never used the term 'character pair', or anything equivalent, when discussing pure types. Again, the reason is simple: all his pure types ever displayed externally were unvarying characteristics. A pure-type plant with respect to height or colour will always produce exact copies of this trait in its offspring. So with no physical cues to go on, Mendel had no need to refer to pairs of anything.

This hints at the most fundamental sense in which we have had Gregor Mendel wrong for the past 100 years. For good empirical reasons, Mendel believed hybrids to be a special case of inheritance. This is because all he could see were external, or phenotypical, effects. From these it seemed self-evident that when breeding occurred between pure types (*CC* and *CC* or *cc* and *cc*), a perfect union took place between the parental elements responsible for any given characteristic. This was why all such unions bred true. But from this perspective, hybrids (*Cc*) represented some kind of unstable deviation from the norm. This is how Mendel put it: 'If it chance that an egg cell unites with a dissimilar pollen cell, we must then assume that between those elements of both cells, which determine opposite characters some sort of compromise is effected.' So unhappy was this compromise, Mendel reasoned, that the two parties took the earliest opportunity to 'liberate themselves from the enforced union when the fertilizing cells are developed'. Some inner force, he assumed, drove the dissimilar units of heredity apart. This was why, when the next generation appeared, about half the plants reverted to parental forms. The crux of the matter is that Mendel erroneously thought hybrid forms to involve completely different physiological mechanics to pure-type forms. He did refer to opposing units of heredity (though not to pairs) and to segregation. But only for the hybrid.

The contention that he was blind to the possibility of gene pairs and of segregation among pure types is reinforced if we consider the seemingly subtle difference between Mendel's notation for the 1:2:1 ratio (**A** + **2Aa** + **a**) and the modern version. Since the early years of the twentieth century, this has been expressed as **AA** + **2Aa** + **aa**, meaning that a quarter of the offspring of a hybrid mating will have two dominant alleles/elements, a half will have one dominant and one recessive, and the final quarter will have two recessive genes.

Double-letter notations are adopted in explicit reference to the gene pairs central to Mendel's Law of Segregation. This simply cannot be read into what Mendel was saying in 1865. He used the single notation of '**A**' or '**a**' in respect of pure types because they were unvarying in the characteristics they displayed. He used the notation '**Aa**' for hybrids as a convenient way of showing that the subsequent generation of plants they produced would reveal that elements for both the dominant and the recessive trait must have been present within them. Rather than being a notation for a gene pair, Mendel's '**Aa**' was no more than a simple means of indicating hybrid status. In sum, whilst Mendel deserves great credit for teasing out the ratio central to modern genetics, the superficial resemblance between his **A** + **2Aa** + **a** and the modern **AA** + **2Aa** + **aa** should no longer be taken as evidence that he gave both genetics and evolutionary theory the crucial insight encapsulated within 'his' Law of Segregation.

In fact, it would be most unreasonable to suppose that he could have done so. The idea of allelic pairs only began to make real sense around 1900—several years after Mendel's death—when scientists had good enough microscopes to detect the existence of the long strands of matter within cell nuclei now called chromosomes. Later, the discovery that gametes hold half the number of chromosomes present in somatic cells inspired the idea that genes, too, might come in matched pairs that segregate during the formation of germ cells. Much more importantly, during the 1910s, in the laboratory of the American Thomas Hunt Morgan, scientists studying inheritance in the fruit fly showed not only the appearance of chance mutations but also how single mutated genes follow the 'Mendelian' principles we accept today.

The transmission of mutated eyes and wings conformed to the Laws of Independent Assortment and Segregation with breathtaking consistency. Before long, new staining techniques had even made it possible for Morgan's team to see the areas on individual chromosomes where particular traits were coded for. Thus, Morgan's experiments proved what his colleagues' mistaken post-1900 readings of Mendel's papers had only suggested, and what Mendel himself could never have known.

Some readers may feel that this is utterly unjust. Knowing what we do, it seems obvious to the modern mind that Mendel's word 'characters' could have been replaced by his term 'elements' to very great advantage. But, as I have stressed, we must put aside what science has learned since 1865. There was absolutely nothing in Mendel's data to suggest that in every case the hereditary units contributed by each parent remain separate entities, conjoined in a gene pair. Nor was there anything to inspire the idea that single genes make an irreplaceable contribution to the formation of the organism. The likelihood is that Mendel shared the then commonly held view that hundreds or evens thousands of hereditary elements are available to specify any given trait. With 'like' elements he therefore felt no need to invoke the principle of segregation. He seems simply to have assumed that there would always be enough passed through to the next generation. In such cases, segregation—in his mind the result of repulsive forces that would not arise between 'like' elements—was simply not an issue.

As it happens, Mendel decided entirely to avoid the question of the numbers of elements involved. He could carry out his breeding experiments perfectly well by just calculating the frequency of the different characters themselves. All he needed to do was record the number of greens to yellows, talls to shorts, wrinkled to smoothed, and so on. In this crucial sense Mendel's was purely a descriptive exercise. It did not matter to him what the genes/elements were doing and how many of them were doing it. This interpretation makes even more sense when we recall that Mendel was not seeking the laws of heredity, but that he was trying to create new species through hybridization. From this point of view the number of elements were of little or no consequence to him. He could quite satisfactorily recognize species by their physical form and this was as close to their genetics as his self-imposed task impelled him to go.

Even on the few occasions in which Mendel did discuss the genetic basis of traits, he preferred to speak of entire cells rather than single elements. Thus, referring to the formation of a pure-type plant, he wrote, 'If the *reproductive cells* be of the same kind and agree with the foundation cell of the mother plant, then the development of the new individual will follow the same kind and agree with the foundation cell of the mother plant.' Not even a whiff of the idea of paired-genes. Furthermore, we have already seen how he described the genetics of hybrid plants. He spoke of the 'egg cell' uniting with a 'dissimilar pollen cell'. Nothing about finite numbers of elements.

Continuing, Mendel explained how a 'compromise is effected' between the 'elements of both cells'. Again, in one of the only passages in which he discusses the elements themselves, there is absolutely no appreciation of their numerical quantity nor of the universality in somatic cells of gene pairs. Insofar as his papers do contain reference to segregation, it is solely in respect of the special case of hybrids and their dissimilar elements. This is made clear in the following passage: 'In the formation of these [germ] cells [of the hybrid] all elements participate in a perfectly free and equal arrangement, whereby only the differing elements are mutually exclusive.' This quotation really clinches Robert Olby's case. It makes abundantly clear that for Mendel the hybrid is atypical because its constituent elements repel at the point at which the sex cells are produced. In Mendel's schema, pure types lack this complexity and for them the provision of segregation is entirely unnecessary.

None of this should be taken as a criticism of Mendel. But it does affirm Olby's suggestion that he cannot accurately be called a Mendelian in the modern meaning of the term. Mendel lacked the evidence to become one, evidence that it would take many more decades to accumulate. To make a point that will become familiar to readers, had Mendel presented his data in 1865 as clear evidence for allelic inheritance it would have been unsupported speculation: not good science.

The Law of Independent Assortment

Precisely the same kind of objection can be raised to the suggestion that Mendel's papers contain 'his' Law of Independent Assortment. As we have seen, in modern genetics this refers to the fact that the entrance or otherwise of any one gene into a gamete does not in any way determine which member of any other gene pairs will join it. The only exclusion that does apply is that its former gene-pair partner will never do so.

Having no conception of gene pairs, this is not a law that Mendel could possibly have devised. As has probably become clear, however, Mendel's great strength lay in observing and tabulating the external effects that would lead others to formulate this law. Mendel carried out a large-scale programme of cross-breeding between hybrids, taking account of a range of different characteristics subject to dominance and recessiveness. What he found enabled him to construct what he called the Law of Combination of Different Characters. Again, note his reference to characters, not elements. As is to be expected, he believed what he had found to be peculiar to hybrids. The underlying principle was that 'hybrids produce egg cells and pollen cells which in equal numbers represent all constant forms which result from the combination of the characters bought together by fertilization'.

What he means by this is that when he crossed hybrids that had two rival forms of, say, flower colour, seed colour, seed smoothness, and height, he obtained roughly equal numbers of all the possible permutations. He got roughly as many tall, purple-flowered plants with green and wrinkled seeds as he got short, white-flowered plants, with yellow and smooth seeds, and so on. To us, it is obvious why this should be so. With each gene's possible entry into a gamete independently subject to the law of chance, it is statistically inevitable that, with large numbers of gametes, all combinations will turn up in roughly equal numbers.

What was not inevitable is that Mendel would somehow intuit why this should be so. Indeed, given the additional ground that would have to be covered before such an intuition was likely, it might be more realistic to say that, given when he was working, it was almost inevitable that Mendel would not come up with the underlying reasons. Mendel's discipline in carrying out all this work, and his high intelligence in seeing that the frequencies of different hybrid combinations were a worthy subject of study, deserve our gratitude and our admiration. But we cannot sensibly go on from this to credit him with formulating an idea that would, and probably could, only emerge after his death.

At this point, it may be helpful to summarize where we have got to so far. First, when Mendel spoke of paired characters, he was not thinking of allelic pairing. This is clear from the way in which he referred only to pairings of characters and not of elements. Second, he believed that hybrids are a special case, even a contravention of nature's laws. This is apparent from his failure to speak of either character or element pairs in relation to pure types and from his description of hybrids as involving an unhappy 'compromise' not required of pure types. Third, although he believed that a process of segregation takes place when the sex cells form, he considered it necessary only for the hybrid cells in which has occurred the 'unnatural' conjoining of 'dissimilar' elements. Pure types are not included in this process because their 'genetics' are assumed to be entirely uncomplicated. Finally, it should be noted, all this is fully compatible with the fact that Mendel was interested in the theory of 'species multiplication through hybridization' and not heredity by itself.

The Mendelian ratios

The Mendelian ratios represent one area in which it might seem to us reasonable to expect Mendel's contemporaries to have got quite excited. Not only did Mendel discover the externally apparent 3:1 ratio, he also revealed the concealed subdivision within the 75 per cent of plants show-ing the dominant characteristic. As he rather neatly explained:

> The ratio of 3:1 . . . resolves itself therefore in all experiments into the ratio of 2:1:1, if the dominant character be differentiated accord-ing to its significance as a hybrid-character or as a parental one. Since the members of the first generation spring directly from the seeds of the hybrids, it is now clear that the hybrids form seeds having one or other of the two differentiating characters, and of these one-half develop again the hybrid form, while the other half yield plants which remain constant and receive the dominant or the recessive characters in equal numbers.

This was not the only time that a ratio close to 3:1 was discovered during the 1860s. Charles Darwin himself recorded a distribution of 2.38:1 from breeding two different varieties of snapdragon. He was unable, however, to make much sense of this recurrent pattern. Mendel had chosen wisely in concentrating for several years on just one species of plant and very few individual characters. Repeatedly encountering the 1:2:1 ratio he could hardly help noticing its importance. And, without doubt, carefully documenting this statistical distribution was the most important legacy Mendel left to the twentieth century. Even on its own it goes a long way to confirming the pivotal role credited to him in the development of what was to become genetics.

There remains the puzzle, however, of why it had so little immediate impact in 1865. Were Mendel's contemporaries foolish, bigoted, or just piqued that a monk should have made the breakthrough? Certainly not. The downside of looking at just one plant species is that it is hard to generalize from it to the rest of the plant and animal kingdoms. Nature is full of diversity and Mendel was well aware that he may have been dealing with a very special case of inheritance. In classification terms, the domesticated pea plant might have been the duck-billed platypus of the allotment garden, something notable for its atypicality rather than the reverse. Mendel would have been encouraged in this wholly mistaken view by the results he obtained in breeding studies involving the bean plant *Phaseolus* that he also reported in 1865. Mendel first explained how the edible pea gave wonderfully clear results:

> In *Pisum* it is known that the characters of the flower and seed-colour present themselves unchanged in the first and further generations, and that the offspring of the hybrids display exclusively the one or the other of the characters of the original stocks.

Then he contrasted this with the transmission of colour in *Phaseolus*. Instead of the two traits under study continuing unaltered into the next generation, in all but one case the 'plants developed flower-colours which were of various grades of purple-red to pale violet. . . [And] the colouring of the seed-coat was no less varied than that of the flowers'. *Phaseolus* did not give Mendel the neat 1:2:1 ratios for which he was hoping. So anybody in the Brno audience who had been impressed by his *Pisum* findings might very reasonably have then lost interest.

Our modern understanding of genetics enables us to see that Mendel did not get the expected results when mating *Phaseolus* because in that plant so many genes contribute to the characteristics in which he was interested. As a result, the underlying 1:2:1 ratio is much harder to identify. Mendel was alert to this possibility, but after 1866 he quite reasonably lost faith in the universality of his ratio. Thereafter, the man who had embraced the monastic life because it afforded him time to indulge his scientific pursuits, became more and more involved in the administration of his monastery. Within a few years he had become Abbot. Thereafter his intellectual powers were mainly directed towards protecting the monastery's assets from the tax-hungry officials of the decrepit Austro-Hungarian Empire.

Mendel and Darwin—a delayed marriage?

Having reviewed what seems to me to be very strong evidence that Mendel never grasped the basic tenets of 'Mendelian' genetics, I now wish to look at the other half of the Mendel legend. This is the idea that had Darwin read Mendel's 'Experiments in plant hybridization', then Darwinism need not have spent almost half a century in a state of limbo neither accepted nor rejected by the Scientific Establishment. The rationale for this belief cuts to the very heart of Darwin's thinking on heredity. As I explain in more detail in Chapter 9, Darwin imagined that reproductive cells begin their lives as buds attached to specific parts of the body. A child's hair, for example, is made up from tiny cells that earlier budded off from its parents' hair cells. In this way, the whole germ cell is the product of buds from every organ and part of the body. Now imagine that a child has inherited blonde hair from its father and black from its mother. If neither colour is dominant over the other, the child will probably develop an intermediate hair colour, say brown. According to Darwin, when buds begin to peel away from the child's brown hair cells, they will contain instructions for brown hair rather than for either black or blonde. The colours have blended in the substance of the 'gemmules' produced.

Darwin's commitment to this theory of blending inheritance was to cause him immense discomfort. During the 1860s his critics seized on blending as a means of demolishing the credibility of Darwinism itself. They claimed that although new hereditary traits might help the individuals carrying them, once these exceptional individuals begin to reproduce with other members of the tribe or herd their special traits will be blended into oblivion. Anti-Darwinians used the analogy of dropping a small amount of dye into a bucket of water, then increasing the amount of water and expecting the colour of the water to get darker and darker even though the dye has become extremely dilute. Yet, because the belief in blending inheritance was so widespread, it was Darwinism rather than the problematic theory of heredity that was placed in the dock.

This is the crisis for Darwinism that conventional accounts suggest that Mendel's ideas could have resolved. Mendel's work with the pea plant directly contradicts the notion of blending inheritance. His second-generation hybrids spoke eloquently of discrete, indivisible units of hereditary material that never underwent a process of blending. No matter how often they were combined with others, Mendel's elements would still regain clear expression.

There can be no doubt that, ultimately, reconciling Mendelism and Darwinism was indeed one of the most fruitful steps in the entire history of biology. Nevertheless, we can be quite sure that had Darwin read Mendel's essays or chatted to him at the industrial exhibition Mendel attended in England in 1862, it would have made no appreciable difference to the out-turn of events at all. For various reasons, the time was not yet ripe for these two sets of ideas to be brought together. For a start, we have already seen that Mendel was strongly committed to demonstrating a rival form of evolutionism based on the idea of hybridization. During the 1860s, Mendel actually had a copy of Darwin's *Origin of Species*. Tellingly, far from seeing its contents as compatible with the findings of his own breeding experiments, his pencilled annotations make clear that he adamantly rejected Darwinism in favour of the Linnaean approach he was himself exploring. Mendel *was* thinking about the origin of species, but he was doing so as part of a tradition largely confined to German-speaking Central Europe. His fascination with hybridization would have made little sense to Darwin whose evolutionary schema (usually) stressed relentless competition, death and non-directionality, rather than the benign, Creation-compatible process of interbreeding that Mendel was striving to demonstrate.

We need also to bear in mind that non-blending theories of heredity were readily available to Darwin during the 1860s and 1870s. Darwin's cousin, the statistician and eugenicist Francis Galton, was explicitly developing an ingenious particulate theory of heredity in which genetic blending was impossible and swamping ceased to be a problem. Why did Darwin ignore Galton? Mostly because he was so deeply immersed in the ancient tradition of seeing germ cells as the buds of body cells. This strongly suggests that, to Darwin, Mendel's evidence would have seemed either inexplicable or a bizarre exception to the general rule. For reasons already given, this is not to imply a lack of openness on Darwin's part. As we have seen, having started with the genetically straightforward edible pea, Mendel was confounded by plants whose chief characteristics were far harder to quantify. We now know that Mendel's difficulty in general-izing his pea-plant findings arose from the more common arrangement whereby characteristics are determined by several genes and not just one pair. This is why most traits in plants and animals are continuous (such as height) and not discrete (such as eye colour). In a pre-genetic age, how-ever, a single plant example of 1:2:1 relationships could hardly have been expected to overturn Darwin's well-entrenched beliefs.

Nor can Mendel be accused of religious bigotry in refusing to face up to the Darwinian direction in which we now think his data points. For the first three decades after the 'rediscovery' of Mendel's ideas, those opposed to Darwinism used Mendel's results and the ideas attributed to him as a stick with which to beat the Darwinians. The British biologist William Bateson, famous for the pivotal part he played in the resurrection of Mendel's work, insisted that Mendelism implied a constancy of hered-itary type over time that was hard to reconcile with progressive evo-lutionary change. This was not an unreasonable position to adopt. As we have seen, the aspect of Mendel's work weighing most heavily on him was his failure, in effect, to produce new plant strains. If any role for religion is to be seen in this, we can only speculate that Mendel might have derived some consolation from his overall failure by reflecting on having unintentionally demolished a central plank of the Darwinian case

Be that as it may, the main point is that pure Mendelism does not, at first glance, fit comfortably with Darwinism: Mendel seemed to have demonstrated the immutability of species, yet Darwinism is predicated on the emergence of new ones. In reality, their much-lamented delayed marriage only became possible well into the twentieth century. A crucial pre-nuptial contribution was made by American work on mutation in the fruit fly and by the evidence gathered by population biologists of the enormous range of variety within single species in nature. The curious notion that Darwinism and Mendelism were kept apart through poor scientific communication alone was devised by Darwinians during the 1930s who wished to counter-attack their critics by belatedly claiming Mendel as one of their own. They were certainly not lacking in effrontery. The realization that Darwinism and Mendelism made a fine matching paiⁱ ɾested on the half-century's intense scientific effort immediately preceding it. Yet, almost immediately after they 'saw the light', some Darwinists began volubly to insist that just given sight of Mendel's papers, Darwin would have had it in one.

Fortunately we are now in a position to be somewhat more circumspect. Once Gregor Mendel is placed back into an intellectual landscape that he would himself recognize, it's clear that he would always have seen *The Origin of Species* as a challenge to his own worldview. For his part, Darwin was also being guided by long-since outdated forms of scientific thought. His lifelong commitment to theories of blending heredity would always have precluded his taking Mendel's results seriously. Seldom can two important scientific thinkers have written at such hopelessly crossed purposes.

Mendel as founding father

When biologists at the turn of the twentieth century read Mendel's 1865 'Experiments in plant hybridization', they read into it fundamental ideas that it simply did not contain. This enabled them to wrench Mendel from a milieu that they did not properly understand and dump him into a context in which he did not fit. Over time, however, it became easy to ignore the passages in Mendel's writings that had looked a little questionable in 1900. Few read Mendel's own essays and even fewer tried to investigate the obscure context in which he had worked. Instead he was catapulted into stardom on secondary evidence, and terms such as 'genius' were used in place of proper historical analysis. The myths that have been allowed to conceal what he was really attempting to do show little sign of vanishing. Mendel's most recent biographer, Robin Henig, takes much of the standard view at face value despite an awareness of more recent scholarship. Indeed, there seems to be an almost universal willingness to skate over the way in which Mendel actually interpreted his results and to ignore the gulf between his worldview and many of the ideas now central to modern genetics.

I have already suggested one reason why Mendel was fashioned into a pioneering genius. Geneticists during the 1930s wished to enrol him as their own prized mascot; a procedure helped by the ease of misinterpreting Mendel's 'Experiments in plant hybridization' as a discussion of genes or 'elements' rather than 'traits' and 'character pairs'. It is also important that three other biologists, Carl Correns, Erich Tschermak, and Hugo de Vries, each claimed around the end of the nineteenth century to have discovered the law of the independent segregation of genes at about the same time. One sociologist has incisively argued that hailing Mendel as the real discoverer may have been intended to defuse a potentially explosive and bitter priority dispute between these three men. But whatever the explanation for Mendel's initial rise to glory, he has maintained his position as a scientific hero because he functions so effectively as a standard bearer for the romantic perception of science. To those who expect their heroes to be unsung in their own time, Mendel stands out as the perfect example. Whereas, for example, Joseph Lister became a Baron and Charles Darwin was awarded the splendour of a state funeral, Mendel died in relative obscurity, his love for science almost entirely unrewarded.

Only now can we appreciate that Mendel actually achieved only a fraction of what is generally thought. Discovering the 1:2:1 ratio took skill, patience, imagination, and self-belief. Even if 'genius' would be going much too far, this deserves to be celebrated as an important breakthrough. Furthermore, it is no discredit to Mendel that modern Mendelian genetics did not emerge, fully formed, from a monastic garden during the middle of the nineteenth century. To think such a feat possible is to overlook how much prior knowledge was required to reach this level of understanding. And to repeat a point I made in relation to John Snow, believing that single individuals are capable of such tremendous accomplishments is also to ignore the fact that science is best viewed as a never-ending, multi-participant marathon, not a series of high-profile relays.

The history of science does include some individuals whose personal efforts are of such quality and so revolutionary as to warrant the label 'genius'. But it is the aggregated contributions of thousands upon thousands of scientific foot-soldiers, junior officers, and men and women of middle rank that account for the great majority of scientific advances. Indeed, understood in his own terms, it is probably fair to conclude that, but for the strategic and tactical objectives of some of those who followed him, Mendel would have remained within this under-appreciated host. His years of admirable dedication were rewarded with posthumous glory but he himself always edged forward with measured steps, remaining blind to where his ideas would one day lead.

WAS JOSEPH LISTER MR CLEAN?

The first method of preventing infection during an operation was developed by Joseph Lister. . . . Lister insisted that the operating theatre was kept clean, that the surgeon wore clean clothes, and that instruments were regularly disinfected.

At first Lister was regarded as an eccentric and nurses resented the extra work that his obsession with cleanliness caused. But deaths from blood poisoning and gangrene were reduced and before he died, Lister's services to medicine were recognized and he was awarded a knighthood. Today the terms, 'Before Lister' and 'After Lister' are used to describe surgery.

BBC's website, *Medicine Through Time*.

What do Jesus Christ, Vincent van Gogh, Albert Einstein, Winston Churchill, and Adolf Hitler all have in common? Although seeming to differ in almost every way, they each share at least one characteristic. Before achieving recognition, fame, or notoriety they are all seen to have spent a period in the wilderness, real or figurative. This image is a very common feature of Western portrayals of the hero or great man. An initially obscure individual performs a remarkably prescient and significant mental or physical feat. They then undergo their wilderness years, before at last their genius is recognized and their names are crowned with immortality. A recurrent theme in biographies—and especially hagiographies—of artists, writers, prophets, dictators, and scientists, it also figures largely in what are explicitly works of fiction.

It is not hard to identify reasons why the image has remained so popular despite seemingly endless reworkings. First, it embodies the

Left: Joseph Lister, first Baron Lister of Lyme Regis (1827–1912).

wish-fulfilling fantasies many of us privately nurse of being 'discovered'. It also reflects an egocentric tendency to overlook parents, teachers, peers, and benefactors when reflecting on our own successes. More importantly, the 'obscurity to immortality' story is perfectly formulated to convey an impression of heroism. Initial rejection straightaway implies both originality and unusual perspicacity. The suggestion of a prolonged period of suffering for the sake of an idea connotes a selfless commitment to the Truth. And an ability to bear the slings and arrows of anonymity and constant repudiation proves that the individual has the requisite mettle to enter the Pantheon of Heroes. Kipling fittingly identifies the ability to 'trust yourself when all men doubt you', as one of the prerequisites of true manhood.

Most of the names adorning the gallery of the Wellcome Institute's History of Medicine library in London's Euston Road immediately conjure up these images of far-seeing genius and altruistic self-sacrifice. We have already seen how John Snow and Gregor Mendel were forced into this romantic schema. But in this particular essay I will focus on the paradigmatic example: the Essex-born surgeon Joseph Lister, son of a Quaker wine merchant and amateur scientist, and a man immortalized by *The Lancet* in 1889 for having 'revolutionized surgery'. Three decades earlier, this *Lancet* editor explained, Joseph Lister's introduction of 'antiseptic' practices had inaugurated a new age in surgical practice. His famous antiseptic carbolic spray symbolized, it went on, both his earnest and pioneering commitment to reducing deaths in hospitals from post-operative infection and the opening up of new possibilities for invasive surgery. The spectacle of insanitary urban hospitals in which infection was spread on the surgeon's instruments from wound to suppurating wound had, it seemed, been banished by the carbolic acid that Lister made famous. Furthermore, the editorial continued, a revolution in understanding the importance of germs in disease begun by Pasteur had found its greatest practical exponent in Joseph Lister.

What *The Lancet* did not point out, however, is that until the 1880s its contributors had rarely had anything good to say about Joseph Lister. As late as 1875 an editorial scathingly wrote him off as unscientific and unworthy of special emphasis. Naturally this was not a problem for his biographers. It served to show that Lister's career had the twin hallmarks of the revolutionary hero: wilderness years followed by international fame. Added to his commanding personality and striking good looks, this produced the perfect subject for a romantic history. The titles of biographies devoted to him, past and present, make clear that this is precisely what he became. For example: *Joseph Lister, Father of Modern Surgery*; *Joseph Lister, the Man Who Made Surgery Safe*; *From Witchcraft to Antisepsis*; *Modern Surgery and its Making: A Tribute to Listerism*; and so forth. In the same vein, the BBC's *Medicine Through Time* website claims that modern surgery traces one of the main roots of its professionalism directly back to Lister's work of the 1860s.

The essentials of this story are clear. In the large metropolitan hospitals of the early to mid-nineteenth century, rates of post-operative death from infection were expected to be as high as 35 per cent and post-amputation fatality rates of 65 per cent were quite typical. Hospital staff, medical and surgical equipment, and fellow patients were all potent sources of infection in the poorly ventilated and seldom-cleaned wards of the period. As our stock images suggest, these wards were over-crowded, dirty, and generally insalubrious. The Scottish surgeon Sir James Young Simpson (Chapter 13) famously remarked that a patient 'laid on an operating-table in one of our surgical hospitals is exposed to more chances of death than the English soldier on the field of Waterloo'.

Hold this impression in the mind. Then picture in split-screen format the classic image Victorian artists sought so hard to recreate of Lister entering the operating theatre as a surgical messiah. A patient lies on a table in the centre of what is, in effect, an amphitheatre. On tiered wooden benches circling the room sit dozens of medical students, colleagues, and adherents. The white-frocked Lister strides towards the patient. He is followed by a procession of helpers and the most prominent of these holds aloft the carbolic-acid sprayer as if he were an incense bearer. Virtually everything about the scene suggests a staged ritual, and, as Lister approaches the patient, the spectres of dirt and disease seem to recede like sin from the world. As a humorous confirmation of this impression, we are told that on one occasion an exceptionally brave student said for all to hear 'And now let us spray!'.

This may be an attractive tableau, but that does not guarantee its accuracy. In this chapter I draw on research—mostly conducted by the historians of medicine Christopher Lawrence, Richard Dixey, and Lindsay Granshaw—that has tested it against primary historical sources. As we will see, what emerges gives us at least some understanding of why, as late as 1875, Lister was dismissed by *The Lancet* as not much more than a self-promoting mediocrity.

Why was Lister out in the cold?

The first problem in fitting Lister to the standard model for heroes is that he was far from alone in trying to reduce post-operative mortality. Over a decade before Lister became a qualified surgeon, public-health reformers had blown the whistle on the conditions inside hospitals. It was generally accepted that airborne 'miasmas', or poisons, were major causes of infection. And, as such, many had joined Florence Nightingale in demanding the closure of the larger hospitals and their replacement with smaller and better-ventilated sanitaria in the countryside where rates of post-operative recovery were already known to be substantially higher. But hospital physicians and surgeons recoiled at this suggestion. It threatened both to lower their public profile and impose severe financial penalties by distancing them from their richest patients. Not least to frustrate Nightingale's proposals, the 1850s saw surgeons introduce a spate of hygiene improvements within the metropolitan hospitals. The regular whitewashing of walls, improved ventilation, the separation of medical and surgical patients, and many other reforms were instituted. Following Continental practice, British hospitals also began to use carbolic acid to disinfect wards.

It was during this drive for greater cleanliness that Joseph Lister became surgeon to the Glasgow Royal Infirmary. In 1865, he processed carbolic acid into a paste that could be applied directly to patients' wounds. Six years later he had developed the Lister trademark carbolic-acid spray and a form of gauze dressing impregnated with carbolic acid. Lister claimed as his inspiration for the elaborate use of carbolic acid the effectiveness with which Louis Pasteur had recently vanquished Felix Pouchet in the debate over spontaneous generation (Chapter 1). He explained that Pasteur's demonstration that putrefaction is caused by airborne organisms was the 'principle' on which his innovations were based. Only the ample use of antiseptics, he insisted, could prevent the same form of putrefaction observed by Pasteur in glass jars occurring in the patient's wounds. At meetings of the British Medical Association in 1867 and 1871, and in a series of books and articles, Lister proclaimed these methods and ideas. He then awaited the expected accolades: these were not generally forth-coming.

The reason for this lack of interest had little to do with either stub-bornness or pique. Writers in *The Lancet* and the *Medical Times and Gazette* rightly observed that Lister's use of carbolic acid was far from innovatory. Some pointed to a prominent *Lancet* editorial of 1864 entitled 'Carbolic acid' in which Dr James Watson explained:

> Several respectable surgeons speak favourably of its use as a lotion—one part of the acid to forty parts of water—in all kinds of fetid ulcer, gangrene and offensive sores . . . Dr. Calvert states that it is the most powerful preventative of putrefaction with which he is acquainted.

Moreover, surgeons criticized Lister's methods after evaluating them according to several entirely rational surgical criteria. First, the critics pointed out, the use of carbolic acid in pastes, sprays, or gauzes was time-consuming and seldom compatible with the busy schedules of metro-politan surgeons. Where dressing wounds had previously been the province of ancillary hospital workers—allowing the hard-pressed surgeons to conduct more operations—Lister's method required constant supervision by surgical staff. In military settings, especially, it was felt that his methods were incompatible with the necessity for speed and simplicity.

Second, and more significantly, many eminent surgeons claimed that their own methods of reducing the rate of post-operative infection were equally if not more effective than Lister's. Sir James Young Simpson, for example, argued that his technique of pinning together the edges of wounds (dubbed 'acupressure') nearly always prevented suppuration. More generally, most of Lister's critics pointed out that maintaining high standards of 'general hygiene' in hospitals had already dramatically reduced mortality rates from post-operative infection. As the eminent English surgeon and pathologist Sir James Paget cumbrously boasted in 1879, 'In the last twenty years, we have had the complete sanitation, as far as the sanitary art has now extended, of all our hospitals in which our patients are contained'.

This was far from being a hollow boast. From the early 1850s onwards, George Callender of St Bartholomew's Hospital in London had been developing a scrupulous regime of hygiene reform within his wards. He described Lister's use of carbolic acid as entirely unnecessary. The lavish use of disinfectant and the isolation of infected patients had, he claimed, spectacularly improved recovery rates. Likewise, George Thomson of Oldham explained in *The Lancet* how his initial enthusiasm for Lister had deteriorated when he discovered that adopting the elaborate carbolic–acid techniques brought no greater benefits to ward hygiene than relying on the extensive use of disinfectants alone.

The key point being made in these criticisms is that Lister focused his attention exclusively on disinfecting the patient's wounds during and after surgery. He not only said little about general hygiene in hospitals, he did little. Even as late as the early 1880s his methods were explicitly contrasted to practices in other hospitals where great emphasis was placed on high levels of cleanliness in wards and operating theatres. Lister's disciples, one surgeon remarked, do not mind if the ward is not 'aesthetically clean', so long as the patient is 'surgically clean'. After a visit to Lister's wards, it was with considerable indignation that another surgeon jotted in his diary in 1871:

> Although great care is evidently taken to carry out the antiseptic treatment so far as dressings are concerned—there is a great want of general cleanliness in the wards—the bed clothes & patients linen are needlessly stained with blood and discharge.

Even in 1883 one of Lister's house surgeons remarked that Lister 'wore an old blue frock-coat for operations, which he had previously worn in the dissecting room. It was stiff and glazed with blood'. News of this getting out would have caused Lister no concern. As he announced to the British Medical Association in 1875, so long as carbolic acid was used he was indifferent to evidence of filth even in his patients' wounds:

> If we take cleanliness in any other sense than antiseptic cleanliness, my patients have the dirtiest wounds and sores in the world. I often keep on the dressings for a week at a time, during which the discharges accumulate . . . and, when the wounds are exposed after such an interval, the altered blood with its various shades of colour conveys often both to the eye and to the nose an idea of anything

rather than cleanliness. Aesthetically they are dirty though surgically clean.

It seems not to have occurred to Lister that antiseptic surgery and general cleanliness would be better than antisepsis alone. The carbolic spray was Lister's hobby-horse and he seems to have decided that adopting any practices used by his fellow surgeons would be an admission of failure. Despite his emphatic concern for them in other respects, this was very bad news for Lister's patients.

Colleagues also accused Lister of violating the new scientific protocols of hospital medicine. At the beginning of the nineteenth century hospitals were little more than repositories for the ailing poor. In most cases, medical professionals were conspicuous by their absence. By the 1850s, however, the large urban hospital had become the fiefdom of a newly prestigious class of surgeons. The resulting concentration of hundreds of patients within a single building, combined with the aspirations of medical professionals to gain the kudos of the 'scientist', created the perfect conditions for the development of medical statistics.

The availability of masses of statistical data relating to cause of death allowed the comparison of different approaches to therapeutics and hygiene for the first time in British medical history. Yet, for all his claims of scientific rigour, Lister bucked this trend. Only once did he publish statistics to adduce the efficacy of his antiseptic surgery. His experience on that occasion did little to encourage repetition. The small sample Lister statistically analysed in 1869 showed an impressive fall in post-operative mortality in his Glasgow wards from 46 per cent to 15 per cent. Unfortunately for Lister, George Callender subsequently published the statistics of 200 consecutive operations that used extensive hygiene precautions in place of Lister's carbolic-acid approach. The results were striking: complete recovery had been achieved in all but six cases. This was a mortality rate just a fifth of that claimed by Lister's now immortalized methods.

Was Joseph Lister in any sense a pioneer?

Other statistics from St Bartholomew's give further evidence of the very limited degree to which Lister can properly be seen as a pioneer. Between 1847 and 1857 St Bartholomew's mortality rate from wound infection was just 15 per cent. In other words, a decade before Lister accepted his first surgical position, St Bartholomew's had already achieved the rate of recovery of which Lister would boast in 1869. Between 1857 and 1867, Bart's mortality rate declined further to 10 per cent; and by 1877 it was at an impressively low 2 per cent. Studies of the relative success of hygiene methods in different wards at the Glasgow Royal Infirmary also found little difference between Lister's wards and those of his sceptical colleagues. In a direct attack on Lister in 1875, one surgeon, James Spence, claimed to have performed 66 amputations without carbolic acid in 3 years with only 3 deaths. If true, this figure contrasts starkly with a Russian visitor's recorded rate of post-amputation mortality among Lister's patients of 17 per cent between 1870 and 1873. In sum, whilst there was undoubtedly a revolution in surgery during the mid-period of the nineteenth century, Lister deserves at most a very modest share of the credit.

At first sight, this seems puzzling. Considering that carbolic acid is a powerful antiseptic, its use should have placed Lister at the top of the surgical merit table. Part of the problem must have lain with the low standard of hygiene in his wards. Given so favourable an environment, some germs invariably evaded the defensive lines of carbolic-acid spray, lotion, and gauze. And, as we have seen, despite widespread criticisms Lister stubbornly refused to take all but the most rudimentary hygiene precautions. In addition, as many contemporary surgeons argued, filling a wound full of caustic chemicals hardly helps promote the natural healing processes. Most surgeons then (as now) preferred to allow the body to heal itself. For this to be successful, the patient's environment must be as clean and as germ-free as is humanly possible. Lister's combination of highly invasive antiseptics plus dirty wards meant his patients often took longer to recover or did not recover at all.

This is not a criticism made possible only by modern knowledge. In 1880, a contemporary of Lister's, the eminent British surgeon Lawson Tait, had his oft-repeated views summarized in *The Lancet* thus:

> For whilst admitting the truth of the germ theory of putrefaction, he maintained that the practice of antiseptic precautions destroyed the healing of wounds, and was accompanied by more constitutional disturbance than careful methods followed without the details of Listerism.

Tait spoke from a position of considerable experience. Over a decade earlier he had used Lister's methods in treating cases of compound fractures. Writing up his results in *The Lancet* he noted that suppuration only seemed to occur in these operations 'when I employed the acid paste exactly as recommended by Mr Lister'.

Lister's relative neglect by surgeons during the 1870s can, therefore, hardly be wondered at. There were too many other members of the profession enjoying spectacular success for his comparatively modest achievements to win him fame. *The Lancet* had every reason to remark in 1875, 'If the special merits of Mr Lister's plan were really as great as they are allowed to be, they should at the expiration of eight or ten years have declared themselves with overwhelming force and certainty'. To Lister's private exasperation, they had not. As a result, in the 1870s, his star seemed to be very much on the wane. Even his 'disciples' began to desert him. Travelling to Scotland in 1873 to witness his erstwhile hero at work, Oldham's George Thomson came back feeling profoundly disabused. In *The Lancet* he condemned Lister's dogmatic commitment to the carbolic-acid spray, and his 'convenient' manner of explaining away failures with his technique as the result of insufficient care in its application. 'The last remnant of my belief in Professor Lister dissipated to the winds', he concluded.

Cresting the wave

How can one explain, then, the fact that Joseph Lister came to be lionized and festooned with awards less than a decade later? To some extent he was being rewarded for decades of genuine commitment to reducing the suffering of his patients. But his refusal to properly compare the level of surgical recovery in his wards with that of his rivals shows that more than altruism was at stake here. Indeed, much of the explanation for Lister's later fame lies in the fact that he was an expert self-promoter. Between 1865 and 1880, he managed to win dozens of loyal supporters (especially in provincial hospitals) and to attract considerable attention abroad. By publishing several books and articles on the subject of antisepsis, Lister also forged a strong link between his name and the concept of reducing post-operative infection. He had, initially, to share this status with

An operation using Lister's carbolic-acid spray.

stronger claimants such as George Callender, but the field opened up considerably in 1878 when Callender died and quickly faded into obscurity. By then Lister had become Professor of Clinical Surgery at King's College London, an appointment supported by most resident physicians and opposed by most resident surgeons. This gave him an even better platform from which to promote his methods and challenge his opponents.

Most important of all, since 1867 Lister had been steadfast in his claim that the use of carbolic sprays and gauzes had been based on Pasteur's 'demonstration' of the role of airborne agents in causing infection. Throughout the 1870s most British surgeons were either uninterested in the ultimate cause of infection or doubtful that Pasteur had provided a complete explanation. But with Pasteur's development of a specific anthrax vaccine in 1881, and the subsequent enormous advances in German bacteriology, Lister could argue that he had at last been vindicated. Before long, pre-eminent German scientists were praising Lister for his prescience and were recommending the use of his spray in military field hospitals. As Lister's stock rose overseas, his British peers were gradually persuaded that they had done this great man a real disservice. During the 1880s they made amends with something approaching abandon.

In trying to understand the enigma of Lister's glorification, one of his contemporaries explained that he had ridden to fame on the 'crest of a wave'. The implications of this metaphor are illuminating. Lister was being swept along by far broader developments in medicine and surgery that began before he became a surgeon and that did not benefit greatly from his practical contributions thereafter. His positioning on the wave's crest was itself highly fortuitous. His decision in 1867 to throw his entire weight behind Pasteur was a genuine gamble. Lister was fortunate that work in Robert Koch's Berlin laboratory during the late 1870s and 1880s led to a craze for discussing and identifying germs as the causal agents of disease. In this context, Lister's not particularly well-grounded intuition was transformed into an apparent case of remarkable foresight. Callender had not embraced germ theory with such single-mindedness. So despite his greater practical success in reducing post-operative mortality, Callender did not qualify for posthumous fame. No doubt many of Lister's erstwhile critics still held to the view that his actual achievements fell a long way short of justifying his celebrity. But by the 1880s, and especially following Lister's elevation to the peerage in 1888, discretion no doubt seemed very much the better part of valour.

Reinvention and renewal

In the early seventeenth century, one of the founding fathers of the scientific method, Francis Bacon, observed that 'Never any knowledge was delivered in the same order it was invented'. The story of Joseph Lister is very much a case in point. Once a key discovery has been made, the meandering route by which it may have been reached seems to some researchers to diminish their achievements. With the benefit of hindsight, there is a strong temptation to doctor the record to present a direct and clear-eyed route to the Truth. Usually these practices amount to little more than minor tamperings. But the construction of scientific heroes often demands a more substantial manipulation. So it proved with Joseph Lister. The ideas he had advanced during the 1860s and 1870s were retrospectively altered in the 1880s. Without these subtle but significant revisions Lister could never have gained such a high degree of posthumous fame.

His greatest rationalizations relate to germ theory. Once it became clear that his early support for this was central to his rising star, he sought to bring that support into line with the ideas of the 1880s. But the record tells a very different story. Look at Lister's earlier writings and it's clear that his understanding of disease and infection is profoundly different from what we believe today. His pre-1880 position regarding germ theory boils down to two basic principles. First, Lister claimed that wound infection is a matter of germs causing already dead tissues to putrefy and release noxious chemicals. He had absolutely no sense that infection can involve living tissues. Instead, Lister extrapolated directly from Pasteur's experiments and imagined infection as involving something like the 'admission of germs into a dead infusion of hay or beef'. 'Healthy tissues', he concluded, 'are capable of preventing the development of these low organisms.'

Second, although Lister identified airborne agents as responsible for wound putrefaction, his concept of what these actually were is far from what he later claimed it to have been. During the 1860s and 1870s, many doctors believed in the germ theory of putrefaction and infection, some to the extent of anticipating modern ideas by arguing that specific diseases and forms of infection are caused by specific forms of disease agent. In contrast, Lister believed that generic microbes float around in the air that only cause specific infections in specific circumstances. Lister's germ was by definition highly 'plastic'. This is totally incompatible with what Louis Pasteur had assumed and Robert Koch and his colleagues subsequently proved. Nor were Lister and his adherents quick to read the writing on the wall. When the German substantiation of modern germ theory was first reported, the Listerians rejected it. Speaking at prestigious medical gatherings, Lister insisted that explanations of disease based on 'absolute morphological characters' are 'entirely untrustworthy'. It is 'not essential', he later repeated, 'to assume the existence of a special virus at all'.

By the early 1880s, however, nobody doubted Koch's remarkable results: his case was so well put forward that it rapidly won the assent of virtually the entire medical and scientific establishments. It was at this juncture that Lister came to appreciate the opportunity that lay open to him: rewrite what he had said on the germ theory for the previous 15 years and present himself as a pioneer of rare foresight. At the end of this process—one that he could hardly have performed without a conscious sense of the distortions he was effecting—Lister felt able to claim that he had been right all along. His strategy was so successful that most of those who have any thoughts on the matter still take him at his own evaluation. Only the original sources tell a different story. In reality, Lister had indeed backed a germ theory of disease. Yet, before he re-wrote history, it was what proved to be the wrong one.

This adjustment in favour of German germ theory was not the only retro-realignment Lister made in the 1880s. His pronouncements on antiseptic surgery also underwent a striking change. As we have seen, in the past he had advocated the use of the carbolic spray and gauze to the exclusion of Callender's more popular techniques of general cleanliness and the isolation of infected patients. Throughout the 1860s and 1870s Lister had shown little interest in the transmission of infection via the solid surfaces of wards and operating theatres. Really known only for one innovation—the carbolic-acid spray—Lister's antiseptic surgery was emphatically not the same approach as keeping wards and operating theatres scrupulously clean. By the early 1880s, however, Lister was puffed up with the plaudits he was receiving from eminent German surgeons.

As his reputation increased, the debates in which he had engaged during the 1870s, and which had unhelpfully polarized approaches to reducing hospital infection, began to seem somewhat trivial and irrelevant. In these circumstances, Lister must have realized that his summary dismissal of George Callender's methods had been utterly misguided. In consequence, during the 1880s he significantly rewrote his own history. Instead of being agnostic towards general cleanliness, if not downright hostile, he re-invented himself as an early advocate. As a result, he was quite inappropriately immortalized not just for inventing an antiseptic spray but also for inaugurating a new age of hygienic surgery. In other words, a large element of his fame was built upon re-assigning to himself the pioneering efforts of his (now dead) professional rivals. Whatever attributes Lister may have lacked, effrontery was not one of them.

Throughout this period, Lister's general strategy was to look back to the achievements of his competitors and interpret their alternative methods as fully consistent with his own approach to 'antiseptic' surgery. It had now become clear that the carbolic spray and disinfectant both worked by destroying potentially deadly bacteria. So Lister expanded the term 'antiseptic' to include anything that served this all-important end. Callender's hygiene reforms became just an extension of an idea Lister had had as early as 1867. Of course, Lister was right in arguing that germ theory—though not the version he had originally supported—showed that there were no theoretical differences between the rival methods of reducing wound infection. His cunning lay in drawing all such methods under his own umbrella term of 'antiseptic' surgery. In this way, he skilfully appropriated credit for successful approaches that he had until recently disavowed. Once again, such was Lister's growing fame during the 1880s that few considered disputing his entitlement to this additional renown.

In time, the term 'aseptic' surgery came to describe methods of preventing wound infection. Ironically the ascendancy of this approach saw the progressive abandonment of Lister's favoured techniques. No longer were wounds to be swamped with powerful germicides. Instead, ever-increasing emphasis was placed on keeping operating theatres, instruments, and dressings sterile. By then, however, Lister's reputation as the leading exponent of all forms of hygiene reform had become so well established as to be virtually unassailable. In the final phase of Lister's career, it would have seemed incredible that this pioneer of germ theory could ever have denied the efficacy of general hygiene precautions.

Even when Lister himself all-but abandoned the use of carbolic acid on wounds few questions were raised. It was seen as a natural development rather than a volte-face. In the same uncritical fashion, it was assumed that he had always been an advocate of general cleanliness and his reputation was allowed to develop untainted. Only the surgeon Lawson Tait stood out against the crowd. Always a strong advocate of general cleanliness, he had an excellent memory and the courage or venom to make public what he knew. Well aware of Lister's strategy of re-invention, he kept up the critical barrage into the 1890s. In *The Lancet* of 1891 he fumed:

To hear the men who wildly proclaimed the necessity of performing surgical operations under a cloud or a stream of potent chemical germicides, declaring now that the essence of their doctrine and their practice was, not the destruction of germs, nor their exclusion, but simple scrupulous cleanliness, is, to say the least of it, startling.

Unfortunately for Tait, the wider world was no longer listening.

THE ORIGIN OF SPECIES BY MEANS OF USE-INHERITANCE

Charles Darwin, whose life spanned much of the nineteenth century, is the most influential biologist to have lived. Not only did he change the course of biological science but he changed for ever how philosophers and theologians conceive of man's place in nature.

John Bowlby, *Charles Darwin: A New Life* (1992).

Lamarck had entertained the obviously fantastic notion that one species might develop into another because of the intense desire and efforts of an individual to better adapt itself to its environment. This, of course, implied the ability to pass on a change in structure to its descendants. Lamarck's ideas were treated with polite irony that relegated them to the sphere of curiosities.

H. R. Hays, *From Ape to Angel* (1964).

Although it is often said that Charles Darwin's ideas initiated a revolution in human thought, it is not always clear which ideas are being referred to. Is it the concept of evolutionary change per se? The possibility of a Godless universe? Or the moral and philosophical implications of our having simian ancestry? There is also the question of whether by attaching Darwin's name to such ideas we are making the assumption that they emerged, fully formed, from one giant intellect. And, if so, are we also presuming that Darwin's theories caused the scales to fall from the eyes of people whose perception of the real world had hitherto been clouded by superstition and dogma?

If we accept the conventional image of Darwin, the answer to each of these questions is likely to be 'yes' in every case. According to this view, the publication of *On the Origin of Species by Means of Natural Selection* in

Left: Charles Darwin (1809–82) towards the end of an intense and wearying life.

1859 triggered a transformation in human consciousness. After Darwin, humanity was forced to retreat from a simplistic conception of itself as the pinnacle of Creation and adopt a view in which fitness for purpose was the only criterion of success. Human beings ceased to be a special case and became just another organism. What Copernicus had already done for our planet in relation to the heavens, Darwin now did for the human race in relation to the rest of the natural world.

Personally committed to such an outlook, the distinguished American zoologist George Gaylord Simpson wrote in 1950, 'The point I want to make now is that all attempts to answer [the question] "Why are we here?" before 1859 are worthless and that we will be better off if we ignore them completely'. Simpson, with most of his colleagues, venerated Darwin and presented the year 1859 as a watershed in human intellectual history: on one side the waters plunged down into ignorance; on the other they flowed forth to enlightenment. To many this remains a picture with enormous appeal; but it comes with a price tag. Accepting it without reservation is to abandon all hope of properly understanding both the man and his work.

In this chapter, I hope to show that if the ever-present temptation of imposing modern understandings of evolution onto Darwin's work is laid aside, he emerges as very much a man of his time. Rather than someone implacably opposed to all that had gone before him, we find that Darwin's thought processes were essentially transitional. His greatest strength lay not in outstanding prescience, but in an extraordinary tenacity coupled with a willingness to admit the apparent power of arguments and evidence that ran strongly counter to his own views. The Darwin that emerges from this re-investigation is no less worthy of our respect. But he is somebody whose career richly illuminates some fundamental aspects of scientific advance.

Charles Darwin: the myth

The popular impression of Darwin is of a reticent, almost hermit-like figure, with a near pathological devotion to natural history. In 1838, this man's fertile brain yielded up an idea that threatened so to transform natural history and ethics that it would bring down on him the wrath of a strongly religious society. Underlying this image is the well-rehearsed, inspirational story of the genius who emerges from inauspicious beginnings ultimately to be crowned with glory.

It begins in 1834 with an exotic flourish. A somewhat directionless young man, Charles Darwin, put off the question of what to do with his life by embarking on a circumnavigation of the globe aboard HMS *Beagle*. During this hazardous voyage his thinking was changed forever by a visit to the Galápagos, a cluster of islands whose obviously volcanic origins meant that they could only have come into being *after* Creation. There he saw how different finches had beaks that were perfectly adapted to their differing island habitats. He also saw how giant tortoises showed no fear, presumably because they have never been hunted by predators. In response to this clear evidence of local adaptation, Darwin began to question the unquestionable: the immutability of species.

Next, there comes the admirable tenacity of a scholar who did not rest until he was convinced that he had found an answer to the most important and pressing questions in natural history. Instead of ignoring facts and opinions that did not seem to square with the ideas he was developing, he repeatedly went out of his way to see if what he was proposing could be falsified. Finally, there is the human dimension of a very private man, hidden away in rural Kent, plagued by panic attacks and daily bouts of vomiting apparently induced by his sense of the enormity of the heresy he was privately nurturing. For 20 years, it is claimed, Darwin wore the burden of Truth like a hair shirt. His revolutionary ideas were so obviously correct that he knew they would inevitably raise a storm if presented to the public. Suffering privately until 1858 he was then galvanized into action by the Welsh naturalist Alfred Russel Wallace, who wrote to him having arrived, decades later, at much the same theoretical conclusions. Such was Darwin's innate decency that, in triggering what can properly be described as his reformation, he ensured that Wallace, too, received due credit. As a closing tableau, we have the image of Darwin the erstwhile heretic heralded as the Newton of natural history, and buried in 1882 with full national honours in Westminster Abbey.

Were an august committee asked to draft a specification for the life most befitting a hero of science, it would be hard-pressed to come up with anything better than the Darwin myth. Given the undoubted importance of his ideas, the dramatic juxtaposition of Darwin the recluse and Darwin the revolutionary alone must have placed him on the fast track to hero status. His life story combines the self-sacrifice of Galileo and the genius of Einstein, all located within the rustic calm of a small hamlet in the picturesque Kent countryside. Brilliant but self-effacing, Darwin was an endearingly English revolutionary.

'Never let the facts get in the way of a good story'

Given the other cases we have looked at so far in this book, I'm sure you will have guessed that there are many problems with this standard account. Its biggest flaw centres on the question of Darwin's originality as an evolutionary theorist. Contrary to the textbook view, none of the concepts from which Darwin pieced together his theory of evolution by natural selection was at all novel. Historians now recognize that the core principles of evolution—struggle for survival, selection, heritability, adaptation, even the appearance of random changes to the hereditary makeup—were fairly common themes in Victorian botany and zoology. Darwin's key contribution lay not in overturning this work, but in recasting it into a more coherent whole.

Even the process of discovery lacks the drama central to the popular account. Although Darwin was impressed by the distribution of species he observed on the *Beagle* voyage, the idea that he experienced a 'Eureka!' moment on the Galápagos Islands is complete myth. He was more impressed by the appearance of flightless birds on the South American mainland than by varieties of finch beak or giant tortoise on the Galápagos. Indicating his lack of interest at the time, Darwin labelled his Galápagos finch specimens so poorly that he later had to approach a London zoologist to work out how different beak specializations might be distributed among the islands. His equally famous giant tortoise specimens are also an invention: the tortoises captured by the *Beagle*'s crew were eventually eaten during the voyage and their remains all thrown overboard.

At a more serious level, recent research has undermined the contrast usually made between Darwin's ideas and supposedly rival evolutionary theories. This work, based largely on Darwin's personal notebooks, shows how deeply he was rooted, throughout his career, in the intellectual trends of the early 1800s. This ground can best be covered by first looking at the evolutionary speculations current in the half-century before Darwin wrote *The Origin*, then at the remarkable extent to which these ideas influenced him, and, finally, at the seemingly impressive evidence that kept most Darwinists locked into these early ideas until at least the third decade of the twentieth century.

Evolution before Darwin

The revamped National Heritage museum located in Charles Darwin's house in Down, in rural Kent, devotes an entire room to examining the evolutionist precursors of its famous occupant. Such honest context-ualization shows a commendable maturing of the curator's role. Rather than being sold a false prospectus, we are enabled properly to appreciate the ideas on which Darwin built. What soon becomes clear from this material is that the notion that humans represent the latest stage in the 'transmutation' of unicellular organisms had been put forward by dozens of naturalists between 1800 and 1859. More specifically, the longstanding idea that Darwin invented the idea of evolution itself is shown to be entirely fictitious. For example, Charles's illustrious grandfather, the poet and doctor Erasmus Darwin, believed that 'all warm-blooded animals have arisen from one living filament'. Thanks to modern research, we now know that during and after the Enlightenment evolutionary ideas of this kind were very far from uncommon.

Thus, an eccentric eighteenth-century Scot, Lord Monboddo, argued that the orang-utan represents an earlier stage in human evolution. During the 1820s and 1830s, the newly founded University College London became notorious as a den for radical believers in human evolution. In France, such ideas were even more energetically and systematically pursued by Jean-Baptiste (Pierre Antoine de Monet) de Lamarck and Étienne Geoffroy Saint-Hilaire. Both won notoriety for advancing detailed 'transmutationist' theories and paid a heavy price for it. As we saw in Chapter 1, there was a crackdown on such expressions of free-thinking after the rise of Napoleon. Their lives were made difficult, and the older and less-versatile Lamarck paid the ignominious penalty of a rapid fall from grace into penury and seeming obscurity.

Each of these theorists was responding to the fact that the traditional view of a static universe had started to crumble. Advances in geology and the discovery of marine fossils at great altitudes suggested a prehistory of enormous length and upheaval. Volcanic activity, ice ages, fossil evidence of extinction, and rock strata representing vast periods of geological time in which there was little or no evidence of life, had convinced most scientists by the 1830s that Genesis was more metaphorical than literal. Reaching this conclusion immediately raised a question of fundamental importance: if change was integral to the history of Earth, how had living organisms managed to avoid extinction and adapt to each new passing geological phase?

Most thinkers preferred to keep God centre stage. Leading naturalists insisted that the Creator had regularly intervened in the life history of our planet. In occasional bursts of creativity, known as 'special creations', he had fully compensated for extinctions caused by changing environmental conditions by producing new forms. Nevertheless, there were always a few prominent thinkers who were dissatisfied with these sorts of accommodation. And in the years after Lamarck's fall from favour, the rapid accumulation of fossils finds were making his ideas seem more and more credible. Fossils of extinct animals that were less elaborate versions of living organisms seemed to them strongly suggestive of a non–supernatural explanation. By the early Victorian period a substantial space for evolutionary theorizing had been opened up. It would never again be closed.

In Germany, ideas of a more rarefied and metaphysical kind fed into the general discussion of species change. The polymath Johann Wolfgang von Goethe was instrumental in producing a school of thought that saw the developmental stages of the embryo as sequentially expressing the hierarchy of forms in nature. German embryologists had noticed that until the later stages of gestation, the human embryo seems to 'recapitulate' many of the developmental stages of 'lower' organisms. Although for these 'naturphilosophers' the development of the embryo did not reflect a real historical progression, their evidence found its way into the most powerful and influential pre-Darwinian book on evolution in which it was given a much more literal spin. In 1844, the Scot Robert Chambers anonymously published his *Vestiges of the Natural History of Creation* in which he traced the law-like emergence of everything from the Universe to mankind in developmental terms, invoking embryology as evidence for an actual evolutionary connection between human beings and rudimentary organisms.

Although Chambers's book was savaged by the British scientific establishment, it was far from being the 'nine-day wonder' that one of its critics predicted. On the contrary, the *Vestiges* had an immense appeal to a proud, ambitious, and self-improving middle class that was losing sympathy with the power of the traditional social elites. Continual middle-class interest ensured that *Vestiges* was reprinted again and again throughout the second half of the nineteenth century, outselling Darwin's *Origin of Species* many times over.

Similarly the English philosopher Herbert Spencer (a man whose own sexual-selection criteria led him to spurn the novelist George Eliot on the grounds that she was too ugly) promoted Lamarckian and embryological notions of human and social evolution from the early 1850s onwards. After 1859, nearly everybody who encountered discussions of evolution did so in his turgid but surprisingly widely read works. Even Darwin relied on Spencer's books more than his own to bring the idea of transmutation to a broad popular audience.

It is only when this first thicket of misconceptions has been cleared away that we can start to tease out which components of Darwin's theory were revolutionary and which were derivative. In doing so, it is crucial to bear in mind a key commonality between the evolutionary ideas of Lamarck, Geoffroy Saint-Hilaire, Chambers, and Spencer. Although they differed in important respects, all shared the assumption that transmutation is the result of forces present within organisms. Evolution was typically presented as the unfolding of an immanent, divine plan that leads organisms towards ever greater progress, complexity, and specialization. This was usually accompanied by the concept of use-inheritance—or the inheritance of acquired characteristics.

Lamarck, for instance, believed that simple organisms were spontaneously generated from lifeless matter under the influence of the 'imponderable' fluids of 'caloric' and 'electricity'. These fluids were internalized by the organism and continuously pushed it towards greater complexity. At the same time, they ensured that any frequently utilized characteristics or behaviours became highly developed. The resultant changes were then passed on to offspring leading to a heightened adaptation to immediate surroundings. By way of example, Lamarckians claimed that the giraffe acquired its long neck because generation after generation of their forebears had stretched for leaves on ever-higher branches of trees. Similarly blacksmiths whose labours had made them exceptionally muscular would produce sons born with a capacity to develop especially large biceps.

Darwin's experiences up to and including the *Beagle* voyage had given him ample opportunity to become fully familiar with such ideas. Not only had his own grandfather been an evolutionist, but his academic mentor at the University of Edinburgh, Robert Grant, had been a committed Lamarckian. During long walks on the beach collecting unusual marine creatures, Grant had delighted in pouring forth his then-seditious ideas to the slightly bemused grandson of Erasmus Darwin. Grant moved to University College London after its establishment in 1827 and won considerable notoriety when he proceeded to give its middle-class entrants the sort of radical, anti-Establishment (and therefore evolutionist) biology for which many of them craved. Our young naturalist had also taken a copy of *Principles of Geology* (1830–3) by the eminent Scottish geologist Charles Lyell on the *Beagle* voyage. During the tedious hours he spent cooped up in his cabin (often avoiding the manic-depressive ship's captain, Rear-Admiral Fitzroy), he had pored long over its critique of Lamarck's evolutionist theories.

By the standards of the day, Charles Darwin's credentials as an evolutionary thinker were impressive. To many of his contemporaries this was to have supped with the Devil, not least because such ideas were redolent of French Revolutionary thought. But if Darwin's more-conservative contemporaries heard the wheels of the tumbrels and the swoosh! of the guillotine blade at the very mention of evolution, there was no denying that evolutionary theory was a powerful and widespread alternative to the orthodox concept of 'special creations'.

Growth and reproduction are as one

Evolutionary theory, then, was hardly new when, during the closing stages of the *Beagle* voyage, Darwin began pondering the natural origins of new species. Over the next 2 years he filled notebook after notebook with thousands of seemingly disconnected thoughts on the themes of evolution, reproduction, and hereditary transmission. By the end of September 1838 he had a semblance of the theory with which we associate him today.

This theory is made up of five main principles. First, to be in a position to procreate, organisms have constantly to fight for survival. Second, on occasion random hereditary variations arise. Third, every so often, one of these slight variations will confer on its bearer an important advantage in the fight for survival. Fourth, this individual and its descendants will procreate more than those organisms lacking the useful acquisition. Fifth, eventually enough of these changes will accumulate in order for a new species to emerge. It is the non-directionality of this model and the randomness of the appearance of variation that distinguish it from the more metaphysical theories of Lamarck, Grant, and Chambers. According to Darwinism, instead of organisms being able in some sense to call forth adaptations from within themselves, environmental pressures favour those that, by chance, prove already better fitted to prevailing circumstances. By the same token, these pressures also act against those less able to cope. The traditional argument that *The Origin* is entirely uncontaminated with earlier evolutionary speculations centres on this crucial distinction.

Detailed investigations of Darwin's notebooks have now made it clear, however, that this presentist interpretation of Darwinism is largely untenable. Darwin never broke away from the core principles of evolutionary ideas that antedated his own. Nor could he have come up with the idea of evolution by natural selection had he not spent the previous years exploring the origin of species with Lamarck and Grant always, figuratively speaking, at his side.

The crucial linkage between Darwin's ideas and those of Grant and Lamarck is a shared developmental view of life in which no firm separation is made between reproduction and growth. Today we know that only extreme levels of radiation and highly noxious chemicals are capable of altering the genetic contents of the ovaries and testes. Normal life events may affect body cells but germ cells are usually much too well insulated to be disturbed. But this distinction between somatic and germ cells is an idea that has enjoyed wide acceptance for less than a century. In contrast, medical and scientific literature from Hippocrates to the early twentieth century are replete with accounts of what were believed to be acquired diseases, such as gout and tuberculosis, being transmitted hereditarily from parent to child. Equally popular were stories of parents losing limbs and subsequently producing legless or armless babies. It was also widely supposed that the thoughts of parents during coitus influenced the character of their progeny. For example, in the absence of any clear notions of genetic predisposition, sex whilst under the influence of alcohol was, of itself, thought certain to guarantee a brood of immoral offspring.

Of course, these ideas sound perverse today. Yet not only did they make sense of family likenesses, but they also accorded with observations of the regeneration of amputated bodily parts in some animals and the nature of propagation in asexually reproducing species. 'One cannot but think', Darwin wrote in 1849, 'that the same power [as heredity] is concerned in producing aphides without fertilisation, and producing, for instance, nails on the amputated stump of a man's fingers, or the tail of a lizard.' All that seemed to be needed to bring these disparate facts into a single explanatory framework was to postulate that variants of the same cells that are involved in bodily growth and repair eventually flow to the ovaries and testes to form germ cells. The production of young became, in effect, pan-generational growth.

Close study of *The Origin* makes it obvious that Darwin considered growth and reproduction to be coextensive in exactly this fashion. His immersion in this ancient paradigm is also crystal clear from the theory of hereditary transmission that he first published in his *The Variation of Animals and Plants under Domestication* (1868). Darwin called his theory the 'provisional hypothesis of Pangenesis'. Its central contention was that every feature of the embryo's body and mind is formed by innumerable specialized hereditary units, or gemmules, that compete for limited 'attachment sites' on the newly created body where they can grow and be expressed. These gemmules were believed to originate in the respective part of the parents' anatomies before budding off and making their way to their reproductive organs.

With this theory, Darwin was able, as he wrote, to 'connect together' and 'render intelligible' an 'astonishing number of isolated facts'. Pangenesis was, in this sense, very good science. But because it is obviously compatible with the now-discredited notion of the inheritance of acquired characteristics, biologists have always preferred to think of Pangenesis as a late retreat forced on Darwin in response to ill-founded criticism. Not so. Modern scholarship has shown that Darwin's belief in the interdependency of growth and reproduction, or heredity and life events, was at the heart of the speculations from which his idea of natural selection ultimately arose in September 1838. Indeed, Pangenesis itself was fully fledged in his notebooks by 1842, some 17 years before *The Origin* was published.

The early notebooks

Darwin's early notebooks have more to reveal. If we decipher the erratic scrawl in those from the period 1836–8, we immediately see how Lamarck's and Grant's ideas profoundly conditioned the way in which Darwin set out to uncover the mechanisms of evolutionary change. At first, we find him toying with the notion that sexually reproducing species have an inbuilt life cycle in much the same way as individual humans. Life begins, in line with a simplistic reading of Lamarck, with a simple organism, or monad, and passes sequentially through the stages of evolutionary development until the onset of species senescence and death. A 'vital' force pervades the entire lineage and determines the course of its development and decline. In other words, at this stage, Darwin was according the environment little or no role.

Then, when Darwin did bring the environment into play, his ideas were even more strongly reminiscent of Grant and Lamarck. He began to argue that changed environmental conditions may 'directly' stimulate the development of new heritable, adaptive traits. 'Condition of every animal is partly due to direct adaptation and partly to hereditary taint', he jotted in late 1837. By this he meant that a species somehow knows what characteristics it needs to cope with a new environment and begins spontaneously to generate them. It was an idea ultimately rooted in religious thought. It was also one very much alien to the notions of selection or random change. 'For instance,' Darwin speculated, 'two wrens, found to haunt two islands—one with one kind of herbage and one with other—might change organization of stomach and hence remain distinct.' The idea of adaptation was there (though this was hardly original), but these two wrens were believed to be capable of adapting themselves to their new diets. They needed to acquire different stomachs to survive. So, as if the wrens' bodies could tell what was required, their digestive systems began to change. Through much of 1837 this schema became Darwin's preferred means of explaining adaptation and change. At this stage at least, he was thinking in ways very different from those now attributed to him.

Rather than slowly fading away, over the following months the ghost of Lamarck continued to dominate Darwin's private thoughts. Although he gradually abandoned the idea of a predetermined species life cycle—of an inevitable development from monad to man—he firmly retained the belief that the environment can directly cause changes to appear in organisms. He now began to argue that monads differentiate into diverse forms throughout the globe as a positive response to environmental effects.

At the forefront of such effects he placed the Lamarckian duo of 'volcanic activity' and 'electricity'. Darwin began to sketch branching patterns in his notebooks, 'trees of life', in which new species budded from the ancestral trunk to form a heterogeneous array of branches and twigs representing new daughter species. At this juncture, and firmly in the context of these classically Lamarckian ruminations, Darwin also made a decisive step towards the theory of natural selection. Extinctions occur, he argued, because not all branches acquire the necessary modifications at the same speed. Those whose inbuilt capacities for physiological or behavioural change are insufficient to allow them to adapt rapidly enough to constantly altering climactic and geological circumstances, go to the wall. In contrast, those that respond rapidly to each change in circumstances continue in the direction of 'progress' and 'perfection'.

Yet, if by 1837 Darwin was discussing the environment as a force selecting only the best adapted forms, such adaptations were not being attributed to random genetic variations. Instead, the driving force was still taken to be environment-responsive modifications that appeared whenever called for. Far from being random, all changes were purposefully acquired and therefore advantageous. Although the initial survival of these self-modifying organisms necessarily depended on the advantages they started with, once faced with environmental change the key determinant became the speed with which they effected their response. As this suggests, Darwin's ideas still bore the goal directedness of earlier evolutionary thinking. In other words, like Grant and Lamarck, he saw the entire phenomenon of evolution as geared towards achieving ever-increasing complexity and sophistication, with the absolute minimum of waste. Evolution was therefore synonymous with progress. Local conditions might fine tune the development of the monad, he argued, but this involved no 'contradiction to constant succession of genera in progress'.

By early 1838, though, the role Darwin was prepared to attribute to the direct effects of the environment in bringing forth new adaptations was rapidly shrinking. He saw in nature such sophistication and complexity that he could not see how an organism could come pre-equipped with the structural potential to meet each new need. In Darwin's attempt to deal with this problem, we encounter a further Lamarckian–Grantian twist. During a period in which he re-read Erasmus Darwin's evolutionist *Zoonomia, or the Laws of Organic Life* (1794–6), rather than seizing on random mutation and environmental selection he opted for the inheritance of acquired characteristics, the concept used by Lamarck, Grant, his own grandfather, and most stockbreeders of the period. Moving on from the neck of the giraffe and the blacksmith's son, in February 1838 he argued that:

> Fish being excessively abundant & tempting the Jaguar to use its feet much in swimming, & every development giving greater vigour to the parent tending to produce effect on offspring . . . All structure either direct effect of habit, or hereditary & combined effect of habit.

Between February and September 1838, Darwin built on this basic model and came to argue that evolution is principally the result of a three-fold process: first, organisms change their behaviour in order to adapt to changing environments; second, over the course of many generations these new behaviours emerge as heritable instincts; and, third, eventually these have the knock-on effect of adaptively modifying the organism's anatomy and physiology.

Only a few months later there occurred the famous 'Malthusian moment' when Darwin re-read Thomas Malthus's *Essay on the Principle of*

Population and—according to many historians—began to see the importance of 'warring of the species', of competition for insufficient resources,
and the inevitability of death to the least competitive. Darwin soon began
to see evolution as capricious and cruel. As he jotted on 28 September
1838, because more offspring are born than there are resources to support
them, this creates 'a force like a hundred thousand wedges trying [to]
force every kind of adapted structure into the gaps in the œconomy of
Nature, or rather forming gaps by thrusting out weaker ones'.

Here, at last, was the theory of the survival of the fittest. At around
about the same time, Darwin also developed more fully an idea that had
cropped up once or twice in his earlier jottings: that the selective process
works on randomly generated modifications. Taking a concept well
known to naturalists and physiologists—the sudden appearance of new
characteristics—Darwin realized that he had the raw material for a model
of evolution that diverged from Erasmus Darwin, Robert Grant, and the
much-abused Jean-Baptiste de Lamarck. Evolution was now the outcome of struggle and chance, not Divine plan, direct adaptations, or an
inner life cycle. Darwin had, in his own words, 'a theory to work by'. Its
heart was pumping weakly, and its umbilical chord was still attached, but
the theory of natural selection was definitely born.

Through the long months in which Darwin had been searching for a
convincing—and conceptually innovative—mechanism for evolutionary
change, the ideas of his evolutionist predecessors provided both map and
crutch. And, contrary to the myth, these were not to be cast aside as his
own theory of natural selection came to the fore. Rather, many of the
core ideas and principles that had informed his speculations between 1836
and late 1838 would always remain with him. Darwin never abandoned
his belief that growth and reproduction are coextensive, nor did he ever
jettison the concept of use-inheritance. Indeed, his final years would see
him beating a retreat to redoubts that he had first constructed in these
early notebooks and never subsequently dismantled. Not until decades
after Darwin's death did evolution by natural selection acquire its modern
dimensions. As will be seen, whilst Darwin carefully nurtured his infant,
he never repudiated its connectedness to the ideas of those on whose
guidance he had so heavily relied.

Darwinism and use-inheritance, 1838–59

The belief that organisms can change their heredity in response to environmental change forms no part of the modern biological orthodoxy. Nor is there any room now for the idea that the degree to which a parent uses an organ will determine whether it is passed down to offspring more highly developed or in a state of atrophy and decline. Darwin, however, clung to such possibilities for the whole of his professional life. In his unpublished 1844 'Essay', the first complete expression of the theory of natural selection, he was still friendly to the notion that the 'direct effects' of the environment can automatically draw forth adaptations from the organism. Even in 1859 he was not prepared to rule out such a mechanism. The only qualification he introduced related to its impact, 'the effect is extremely small in the case of animals, but perhaps rather more in that of plants'.

Much more importantly, Darwin never once shed his earlier commitment to the theory of use-inheritance. The causal role he attributed to it is made clear in *The Origin*. 'I think there can be little doubt that use in our domestic animals strengthens and enlarges certain parts, and disuse diminishes them; and that such modifications are inherited', he explained in the first edition. As evidence of the 'effects of disuse' he cited the diminutive wings of flightless birds such as the logger-headed duck of South America and the domestic Aylesbury duck. By his account, their wings were small because their ancestors had no longer needed to fly. This in itself had altered their heredity. The 'Darwinian' notion that breeders of Aylesbury ducks had chosen to breed selectively from those ducks that had less chance of escaping or that most rapidly put on weight was not put forward.

Where natural selection was brought into play, it was often assigned no more than a supporting role. In explaining the blindness of a South American burrowing rodent, Darwin wrote, 'As eyes are certainly not indispensable to animals with subterranean habits, a reduction in their size with the adhesion of the eyelids and growth of fur over them, might in such case be an advantage; and if so, natural selection would constantly aid the effects of disuse'. Likewise, in his *The Descent of Man, and Selection in Relation to Sex* (1871), Darwin explained the strength of 'muscles serving to express different emotions' and 'the increased size of the brain' among white Europeans as being due to their greater use of these attributes in comparison with other races and civilizations.

Many of Darwin's latter-day supporters have made attempts to excuse his continued loyalty to use-inheritance. Usually it is seen as a precautionary strategy on Darwin's part that gave a fall-back position should attacks on natural selection become overwhelming. But this sort of rationalization will not stand up to even the most cursory examination of his notebooks from the 1830s and 1840s. Darwin's commitment to the idea of use-inheritance was entirely consistent with his fundamental belief that there is no means of separating growth and reproduction. His long-standing genetic theory of Pangenesis provided a well-developed theoretical explanation for the inheritance of acquired characteristics. And in 1859, Darwin was still very doubtful that natural selection alone could bring about adaptive modifications of a kind and scale that would be required, for example, in the stomachs of the two wrens with different food sources he had first considered in 1837. Had he had no alternative explanation, we can speculate that Darwin's commitment to natural selection might have been total. But use-inheritance was just such an alternative and he was always happy to use it.

The indirect effects of the environment

We can now turn to Darwin's notion of 'spontaneous modifications', the random genetic changes taken to distinguish his evolutionist thought from Lamarckism. The first thing to appreciate is that Darwin did not anticipate anything like the modern understanding of how genetic variations arise. It is now known that most mutations occur because of mistakes made in the copying of genetic material. Very occasionally sections of DNA become dislodged and find a new home on their chromosome. This may sufficiently shuffle up the DNA sequence for the functioning of the genes involved to be changed. In other cases, during replication, individual letters in the genetic code are accidentally switched. Again, this may mean that the section of DNA starts coding for different proteins. Sickle-cell anaemia, for example, is caused by a single-letter change in the gene coding for red blood cells: a tiny modification that has major consequences. Crucially, in nearly all such instances, the origin of copying errors has nothing at all to do with the wider environment in which the organism lives. The sickle-cell anaemia mutation is likely to have occurred hundreds of times in human history, throughout the world and in all manner of environmental contexts. Only in malarial regions, however, where the sickle-like shape of the red blood cells make them less susceptible to the malarial parasite, has natural selection favoured its survival.

An enduring myth about Darwin's attitude to what we now know as 'genetics' is that, with impressive scientific forbearance, he placed the question of the ultimate causes of hereditary variation in a 'black box' to be explored only when more information became available. This claim is groundless. By the early 1840s, Darwin already felt he knew the route by which new variations are generated. From then on, when he spoke of the appearance of random variations he used a well-honed phrase, 'the indirect effects of the environment'. He never questioned that the 'conditions of life' somehow provide the raw material for the evolution of new species. As he wrote in *The Origin*:

> Indirectly, as already remarked, they [i.e. the conditions of life] seem to play an important part in affecting the reproductive system, and in thus inducing variability; and natural selection will then accumulate all profitable variations, however slight, until they become plainly developed and appreciable by us.

Or, as he wrote in his 1844 'Essay', stressing his belief that the 'conditions of domestication' themselves enhance the likelihood of new varieties occurring:

> It would appear as if the reproductive powers failed in their ordinary function of producing new organic beings closely like their parents; and as if the entire organisation of the embryo, under domestication, became in a slight degree plastic.

External conditions affecting the 'reproductive system' and increasing the 'plasticity' of the embryo are the key concepts here, themes that are repeated throughout *On the Origin of Species* and *The Variation of Animals and Plants Under Domestication*. This emphasis on the causal role of the external environment shows that Darwin was thinking in fundamentally different ways from modern biologists. So to what was he referring?

For Darwin, the laws of heredity ordinarily served to ensure the production of exact replicas of parental types. But, he argued, there is a dynamic relationship between extraneous circumstances and the laws of growth that can lead to the subversion of the normal process of hereditary transmission and the consequent production of spontaneous variations. Recall that for Darwin the 'reproductive system' was a term that applied to the mechanisms of growth *and* conception. Darwin now reasoned that if during the process of development an embryo is exposed to strange environmental pressures its normal pathway to maturity may be disrupted. As a result the embryo might acquire unusual modifications to its physical or mental make-up. Only chance will dictate whether or not these are useful. Either way, the modifications will involve the production of new cells, which will eventually bud into gemmules and the instructions for the new characteristics will be sent forth into the next generation.

As Darwin put it, 'the reproductive system is eminently susceptible to changes in the conditions of life; and to this system being functionally disturbed in the parents, I chiefly attribute the varying plastic condition of the offspring'. The modifications didn't occur during the process of forming germ cells. Rather, they arose as a consequence of the environmental conditions in which the embryo matured. And the changes wrought to its makeup were exposed to the forces of natural selection in its own lifetime and during those of its offspring.

If Darwin was vague on the physiological details of how all this takes place, he was perfectly at ease in suggesting that the environment in some way elicits heritable physiological or behavioural changes in the developing organism. His entirely conventional nineteenth-century conception of the relationship between growth and reproduction had led him down cognitive pathways greatly at variance with the modern conception of random mutations being the product of the internal mechanisms by which strands of DNA are copied. The irony of the situation lies in the fact that it is to the latter alone that the term 'Darwinian' is now applied.

Pride and progress

The same pattern is to be found in Darwin's treatment of the notion of progress in evolution. It now seems almost axiomatic that much of the brilliance of Darwinism lies in its presenting an open-ended model of evolutionary change. Variation is random and the only yardstick for success is how many offspring an organism manages to produce before succumbing to one of the myriad causes of death nature provides. In view of this, *The Origin* has always seemed to represent a major advance on the 'naively' purposeful systems created by Jean-Baptiste de Lamarck, Robert Grant, and Robert Chambers in which—to nobody's great surprise— human beings are seen as the apotheosis of all evolutionary processes.

Yet this, too, is an interpretation that requires major qualification. Once again, if we look closely enough at the relevant texts, we find not Darwin the iconoclast but Darwin the man of his time. Although he rejected the idea that we all carry within us a blueprint for continued progress, his theory of evolution by natural selection did have a strong progressive component. In *The Origin* he wrote: 'the more recent forms must, on my theory, be higher than the more ancient; for each new species is formed by having had some advantage in the struggle for life over other and preceding forms'. Now this does not necessarily mean that Darwin was loathe to forsake the comfortable and comforting idea of progress in nature. Elsewhere, however, he made his meaning clearer: 'Man selects only for his own good; Nature only for that of the being she tends.' And towards the end of the book: 'As natural selection works solely by and for the good of each being, all corporeal and mental endow-ments will tend to progress towards perfection.'

It is arguable that Darwin included these passages as a prudent means of de-fanging the harsh reality he was presenting to the world. Many of his readers, especially those weaned on the soaring optimism of Chambers and Herbert Spencer, were unprepared to accept the brutal worldview *The Origin* implied. And it is almost certain that Darwin realized that forcing readers to look into the abyss of a nature 'red in tooth and claw', and accept the idea of an essentially purposeless universe, were hardly likely to bring him popular success. Yet it seems improbable that this was the sole reason that he introduced the principle of progress. As the

identical concept appears in Darwin's very earliest evolutionist speculations, it is unlikely that the latter references were mere window-dressing.
There is a major gulf between Darwin and writers such as Lamarck,
Erasmus Darwin, Grant, and Spencer in that whereas they saw the agent
of improvement as largely internal, Charles Darwin mostly placed it in
the external environment. But despite this innovation, he remained very
much Erasmus Darwin's grandson: the source of pressure may have been
external, but the direction of the resultant flow was remorselessly in the
direction of greater complexity and progress. The reception of their
respective ideas was, however, very different.

Recall that in Erasmus's day, progressive evolutionary theories were
considered seditious. Since then, however, profound societal changes had
rendered the idea of progress entirely respectable. Victorian Britain's
unprecedented rate of economic and technological advance had fostered
a passionate belief in the inevitability of further progress. And an almost
visceral state of national optimism had come to pervade all forms of
Victorian social and intellectual life. This is the Britain of the Great
Exhibition, proud possessor of an Empire of unprecedented size on which
the Sun never set. The resulting cultural milieu ensured that only the
most conscientiously pessimistic Victorian scientist could resist placing
progress at the core of their evolutionary models. Always (over) alert to
the nuances of public opinion, Darwin was in this sense, too, very much a
man of his time.

Putting Darwinism on hold

In 1867, a young man from Boston named Henry Adams travelled to the
United Kingdom and recorded his impressions of a period convulsed by
debates over Darwinism and man's place in nature. Writing in his autobiography in the third person, he recalled:

> Darwin hunted for the vestiges of Natural Selection, and Adams
> followed him, although he cared nothing about Selection, unless for
> the indirect amusement of upsetting curates. He felt, like nine men
> in ten, an instinctive belief in Evolution, but he felt no more concern
> in Natural Selection than in unnatural Selection.

This passage beautifully conveys one of the most important facets of the reception of Darwin's theory of natural selection. Within a decade most scientists, a large proportion of laymen, even many senior churchmen, had gladly embraced the essential 'truth' of human evolution. Even though the fossil record was uncomfortably patchy, Darwin's ideas enjoyed too much high-level scientific support and resolved too many outstanding scientific problems, to be as casually dismissed as had the evolutionary theories of Erasmus Darwin, Lamarck, and Chambers.

Nevertheless, although the idea of 'evolution' had burst through into open country, the concept of 'natural selection' was still dug-in on the beaches, pinned down by a blistering barrage of enemy fire. The reasons for the discrepancy in popularity between natural selection and evolution *per se* were partly cultural. Darwin himself only partially appreciated the non-directionality implied by *The Origin*. For many of those who more clearly glimpsed the relentless struggle being waged within every crevice and rock pool of the natural world his ideas suggested, the tidier and more obviously progressive evolutionism of Chambers and Spencer was much more appealing. To the astronomer Sir John Herschel, the idea of natural selection amounted to what he derogatorily called the 'law of higgledy-piggeldy'. Similarly many of Darwin's more-pious readers wished to uphold neo-Lamarckian theories in which God's beneficent hand was somewhat easier to make out.

But the majority of readers were simply not that discerning. Most would have struggled to identify how Darwin's schema differed from the neo-Lamarckian ideas offered by Spencer and Chambers. 'Darwinism' became synonymous with just about any form of evolutionism that could possibly be imagined. Partly because of this lack of discernment and partly because, as we have seen, Darwin was far from dismissive of his ideas, by the year 1900 Lamarck was very much back in business. The idea that evolution follows a preordained course always enjoyed great popularity as did the notion that improvements gained within one generation may be passed down to offspring. Both concepts had much to recommend themselves to a society that was getting stronger and richer all the time and wished to feel that the honeymoon would never end. Thinking very much in terms of the modern conception of Darwin's ideas, Julian Huxley—the grandson of Darwin's 'bulldog' Thomas H.

Huxley—would later write of this period as witnessing the 'eclipse of Darwinism'.

More important, though, than cultural context in explaining this 'eclipse' were several empirical factors that seemed to make alternative evolutionary theories far more scientifically credible. From as early as 1859, Charles Darwin was frustrated by Thomas Huxley's failure to endorse the theory of natural selection with any genuine conviction. Huxley was persuaded by the anatomical similarities between human beings and the 'lower' apes that evolution takes place, but he remained largely agnostic on the mechanisms responsible. With a typically self-deprecating air, Darwin had himself remarked to Huxley in 1859, '[my theory] is a mere rag of an hypothesis with as many flaws & holes as sound parts'. By the mid-1860s, the most important of these 'holes' soon threatened to destroy the shaky edifice that Darwin had so painstakingly constructed. When he wrote *The Origin* he knew that the gradual evolution of mankind by the random mechanisms that he was invoking would require possibly thousands of millions of years. In 1859, he set aside an enormous 300 million years since the extinction of the dinosaurs alone. At this stage, Darwin was reassured by his geological training that Earth was at least as old as he required it to be. 'We have almost unlimited time', he declared in 1858. Cue the physicists.

In 1865, the Scottish physicist William Thomson (later Lord Kelvin), dropped a bombshell. His argument was based on a simple application of the laws of thermodynamics. Assuming that Earth was once a mass of molten rock, one could very roughly estimate the time it would take for its surface to cool down sufficiently to permit the emergence of life. Kelvin did so and announced that only 100 million years had elapsed since crustal condensation had rendered the globe habitable. Panicked, Darwin hurriedly asked his mathematical son, George, to recalculate Kelvin's figures. Unless the planet is many times older, he intimated, 'my views w^d be wrong'. George's reply brought him no joy. And such was the prestige of physics in general and Kelvin in particular, most Darwinians felt that they had to come to heel and adapt themselves to the new conditions Kelvin had created. We now know that Lord Kelvin was wrong for reasons that would emerge only through the discovery of radiation around the year 1900. But in the nineteenth-century context, support for

William Thomson, first Baron Kelvin of Largs (1824–1907).

natural selection had suffered a giant setback from a seemingly unimpeachable scientific source.

Unfortunately for Darwin, the Earth's age was only one of the two
major storm fronts that threatened to converge and sink the theory of
natural selection during the late 1860s. The other arose in 1867 in a
widely read article by the English engineer Fleeming Jenkin. Jenkin's
basic point was that, given the theories of 'blending' inheritance to which
Darwin and most of his contemporaries subscribed, any new heritable
modification would soon be diluted and eventually entirely obliterated

when combined in reproduction with the rest of the population. Imagine a modification that produced a purple-furred rabbit in a population of white rabbits. Even if the new colour conferred some immense adaptive advantage on the individual affected, as it would have only white-furred mates with whom to interbreed, whatever gave rise to the purple fur would soon be diluted out of existence. This was a devastating argument that could be effectively combated only after 1900 with the development of a non-blending theory of heredity—what the English geneticist and statistician R. A. (later Sir Ronald) Fisher came to term 'The genetical theory of natural selection'. Back in 1867, however, Jenkin's critique cost Darwin a great deal of sleep. Whilst Jenkin was not attacking the concept of evolution per se, it was quite apparent that natural selection's suggested contribution to evolution had been dealt another heavy blow. As a result, it fell even further in the estimation of leading scientists.

What made matters worse was that Darwin fought his battles from a distance and with what looked like a lack of conviction. Easily pushed into bouts of extreme anxiety, vomiting, and chronic diarrhoea, his preference was for his able lieutenants—Thomas Huxley and the eminent botanist Joseph Hooker (both of whom we shall meet in Chapter 10)—to wage war for him instead. From his rural retreat in Kent, he wrote submissive letters encouraging them to divert the fire away from him at the same time as defending his intellectual credibility. Writing to Hooker in 1860—after Hooker had debated publicly in favour of human evolution at the famous meeting in Oxford of the British Association for the Advancement of Science—Darwin thanked him, 'It is something unintelligible to me how anyone can argue in public . . . I am glad I was not in Oxford, for I should have been overwhelmed'. This strategy of deprecating himself in order to buy the continued loyalty of his allies was part of Darwin's stock in trade. It may sound like insufferable obsequiousness, but this was a man so genuinely afraid of public appearances that he had a mirror fitted in his study that gave him advance warning of anyone walking up his drive.

There was, however, one way in which Darwin could still seek directly to influence the debate: by writing further editions of *The Origin*. In the summer of 1871 and the winter of 1872 he worked on the final edition of his great work. During this period he sought to perform emergency repairs to an ailing theory. Although Darwin never fully accepted Kelvin's calculations (he called Kelvin an 'odious spectre'), he did concede that evolution had to be squeezed into a far shorter timescale. The effectiveness of Jenkin's critique had also convinced him that the theoretical danger of useful adaptations being swamped out could not be ignored. The fifth edition of *The Origin* had already made some concessions to these critics, the sixth and final edition would make many more.

Darwin sought to meet the challenge in two ways. First, by arguing that in certain periods the environment (indirectly) causes a considerable increase in the rate at which new modifications occur. Contradicting an earlier opinion he had expressed, Darwin explained that:

> The world at a very early period was subjected to more rapid and violent changes in its physical conditions than are now occurring; and such changes would have tended to induce changes at a corresponding rate in the organisms which then existed.

Darwin also insisted that his critics had underestimated 'the frequency and importance of modifications due to spontaneous variability'. This important—if unsubstantiated—claim permitted him to argue (1) that spontaneous modifications will arise sufficiently often for the danger of blending to be avoided; and (2) that evolution by natural selection is reconcilable with a young Earth.

But most of the changes Darwin made to the final editions of *The Origin* involved retreating to the quasi-Lamarckian positions he had developed in 1837 and 1838. This retreat would continue throughout his final years. In 1880, two years before his death, he wrote to Alfred Russel Wallace: 'It is impossible to urge too often that the selection from a single varying individual or of a single varying organ will not suffice.' Accordingly, his second and most important modification to *The Origin* was to considerably enlarge the role ascribed to use-inheritance. As natural selection alone was extremely wasteful and time-consuming, it seemed logical to argue that evolution had been accelerated by the inheritance of any acquired characteristics that were of survival and/or procreative value. Darwin found that for every blow received to his theory of natural selection he could compensate by adding further emphasis to the notion of use-inheritance. There were limits to how far he could push this strategy if he were to avoid entirely disappearing into the shadow of Lamarck. The indirect effects of the environment therefore remained paramount even in the final edition of *The Origin*; but it was only by stressing use-inheritance that Darwin was able to embrace death feeling that he had escaped the bear-trap Kelvin had dug for him.

To cope with Fleeming Jenkin, Darwin adopted another approach. In 1867, Jenkin himself observed that swamping need not take place if similar modifications occurred virtually simultaneously within a single breeding population. Here Darwin made his greatest concession of all to earlier ideas:

> There must be some efficient cause for each slight individual differ-ence as well as for more strongly marked variations which occasion-ally arise; and if the known cause were to act persistently, it is almost certain that all the individuals of the species would be similarly modified.

In this remarkable passage Darwin was still positing an external cause of change, but variation generated by environmental conditions has ceased to be *random*. Certain environmental conditions call forth particular modifications simultaneously in many members of the species. In an attenuated form, then, the idea of the environment 'directly' engendering adaptation had been dragged once more to the fore. Pushed much further and this argument would have wrecked the theoretical system that Darwin had spent more than two decades constructing. One is therefore left wondering what would have been left of natural selection had Darwin not died when he did.

Context and contingency

To those biologists who have delved beneath the myth outlined at the beginning of this essay, the changes Darwin made to the later editions of *The Origin* are a source of profound disappointment, even embarrass-ment. For example, the Oxford biologist Cyril Darlington complained in 1953 that Darwin 'panicked and ran straight into the opposite camp . . . Lamarck became a posthumous Darwinian'. In his recent *Something Like a Whale* (2000), the British geneticist Steve Jones has opted for the more kindly judgement that Darwin was 'worried' by his 'ignorance' of the subject of heredity, and this 'led him, in his later years, to complicate and confuse his ideas'. Yet, to impute a sense of ignorance to Darwin is almost certainly a mistake arising out of the wealth of modern knowledge. It is unlikely to be something Darwin himself keenly felt. And, as has already been suggested, his recourse to Lamarckism was not undertaken with the profound reservations Jones implies. For Lamarckism never represented 'the enemy' to Charles Darwin.

Not only had quasi-Lamarckian ideas given birth to the theory of natural selection, but the concepts of use–inheritance and progress always remained integral parts of Darwin's evolutionary system. Between 1865 and 1882 he responded to several cogent criticisms of *The Origin* by retreating to fall-back positions he had never really given up. Nor was retrenchment in any sense a flight from reason or a clumsy, ad-hoc attempt to bolster his theory. Entirely lacking the crucial insights provided by modern genetic theory, Darwin was behaving in a thoroughly empirical fashion in qualifying the first edition of *The Origin*. Unfortunately for him, until the 1920s and 1930s the evidence just did not tell in his favour. Between 1859 and 1882, as he himself said, there were quite simply as many 'flaws & holes' as 'sound parts' to his theory.

'A IS FOR APE, B IS FOR BIBLE'

Science, religion, and melodrama

> The Bishop displayed his ignorance of Darwin's ideas and sneeringly asked Huxley if he traced his descent from a monkey through his grandfather or his grandmother. Huxley demolished 'Soapy Sam', as his enemies called him, with cold logic.
>
> H. R. Hays, *From Ape to Angel* (1964).

> Huxley and Hooker annihilated Wilberforce's position at the Oxford debate and continued spreading what was tantamount to a gospel of evolution.
>
> *Encyclopaedia Britannica* (1992).

In 1994 and 1995, the University of Oxford hosted two academic bun fights that went under the deceptively grandiose heading 'The Science versus Religion Debates'. Hundreds of listeners crowded into a modern lecture theatre to see a series of eminent scientific guest speakers pitted against several less well-known representatives of revealed religion. Oxford's high-priest of evolution, Richard Dawkins, was a participant both years and in 1994 I watched as he scythed through the opposite panel's arguments with obvious relish. At the time of the second debate, I happened to be reading a biography of Charles Darwin written by the historians Adrian Desmond and James Moore. Again, I watched Dawkins's by now almost-formulaic humiliation of the theological faction. This time, however, a vivid image formed in my mind of an event that had taken place only half a mile away almost a century and a half before. No doubt an overheated imagination was at work, but I could almost see Thomas H. Huxley, Darwin's 'bulldog', and the Bishop of Oxford, 'Soapy' Samuel Wilberforce, sparring on the stage, a scene made all the more dramatic by the collapse of a crinolined lady and a livid

Left: Thomas Huxley (1825–95) lecturing on the gorilla; Samuel Wilberforce (1805–73), Bishop of Oxford.

Rear-Admiral Robert Fitzroy, Darwin's captain on the *Beagle* voyage, leaping to his feet grasping the Bible between his weathered hands, and bellowing at the top of his lungs 'believe God not man!' and 'Darwin is a viper!'.

Assuming that Dawkins accepts the traditional account of what went on that night, he would have good reason to be content with this implied comparison to Thomas Huxley. For what happened in Oxford on the night of Saturday 30 June 1860 is generally considered to have been of the most profound importance in shaping the future relationship between science and religion. To set the 1860 Oxford debate in context we have to imagine that only 8 months have passed since the first publication of Darwin's *On the Origin of Species*. The debate about man's place in nature is in the air but few have had the stomach or the audacity to voice it publicly. Tonight, however, is going to be different. A British-born American, John Draper, has agreed to address the prestigious British Association for the Advancement of Science, meeting in Oxford, on 'Darwinism and human society'. Arrayed before the gutsy American, and crowded beneath the Gothic Revival arches of the newly built science museum, are the leaders of British scientific and clerical opinion. Unsurprisingly the clergy are out in force. Oxford University's then primary function was to train clergymen. And these same clerics, viscerally opposed to Darwinism, can be imagined as bracing themselves to make a heroic stand. There is to be no more appeasement.

Draper delivers his paper without much verve and with the minimum of offence. Then the heavyweights, minds and sinews taut for battle, take the stage. There now unfolds one of the most celebrated exchanges of repartee in British history. With well-honed sarcasm the stridently anti-Darwinian Bishop Wilberforce asks Huxley whether it is 'on his grandfather's or grandmother's side that he is descended from an ape'. Huxley whispers beneath his breath, 'God hath delivered him into my hands!', and then responds with imperious contempt:

> If then the question is put to me would I rather have a miserable ape for a grandfather or a man highly endowed by nature and possessed of great means and influence and yet employs these faculties and that influence for the mere purpose of introducing ridicule into a grave scientific discussion I unhesitatingly affirm my preference for the ape.

Writing to his friend Dr Dyster shortly after, Huxley explains how this devastating rejoinder caused 'inextinguishable laughter' in the chamber. With the help of his fellow Darwinians John Lubbock and Joseph Hooker, he enlarges, 'We shut up the bishop & his laity'.

Savouring their victory, the Darwinians later reflected on how in Oxford the first blast of the trumpet against the organized trickery of revealed religion had been sounded. Science was finally winning its autonomy from the Established Church. After this night, they claimed, it would become harder than ever to overlook the fatuity of the Biblical explanation of human origins. And in the years that followed, religion would shrink before the majestic power of Darwinian science, bludgeoned by Darwin's loyal crusaders into the few remaining regions of human mystery. The night of Saturday 30 June 1860 was, in short, a landmark in the victory of scientific reason over faith and obfuscation. At least, that is how Huxley & Co. saw it. Were they right? Well, not exactly.

The epic confrontation of June 1860 has been described as the most famous battle of the nineteenth century after Waterloo. It may, therefore, come as a surprise to learn that scarcely an element of the legendary exchange between Huxley and Wilberforce has not been exaggerated to the point of absurdity. As with the other myths examined in this book, for essentially strategic reasons the Oxford debate has been invested with profound symbolic meaning. In the process, the Truth was a very early casualty. The legend of Huxley's slaying of Wilberforce has been immortalized largely because explicit confrontations between science and religion allow scientific propagandists to show science in the best of all possible lights. In short, religion has for long been the perfect foil for the militant and zealous scientist. Theologians, they point out, are committed to an unfalsifiable collection of core ideas that they must accept, as Tennyson put it, 'by faith, and faith alone'. In contrast, scientists attest— often legitimately—to a willingness to forsake even their most cherished convictions if the evidence so demands: science permits free enquiry unburdened of dogma and cant.

Yet, there is a real price to pay for accepting this over-simplified view. Presenting science and religion as natural antagonists obscures the real and fascinating complexity of their relationship in the history of science. Dismantling the enduring myth of the Oxford debate enables us to explore this relationship and to show how easily—and from what meagre resources—the gospel of the Progress of Science has sometimes been constructed.

Wilberforce's Waterloo?

Few accounts of the June 1860 debate have resisted the temptation to ham-up its dramatic side. The fainting of a crinolined Lady Brewster and Rear-Admiral Fitzroy's wild fulminations are both cases in point. In fact, even these episodes should not really surprise us. The heat of the chamber and the highly circumscribed role and corsetry imposed on women in Victorian society made it likely that at least one of their number would, unintentionally or by design, slip into unconsciousness. Similarly it was only 5 years later that Fitzroy's notorious mental instability led him to take his own life, slitting his throat as his uncle, Lord Castlereagh, had done 38 years earlier. But these are minor embellishments to the traditional story. By far the most remarkable feature of the Oxford debate—as compared with the myth that later grew around it—is that *all* the main protagonists left the Oxford Museum well satisfied with their performances and convinced that they had personally carried the day.

Thus, Wilberforce, a seasoned and skilful debater, was assured that his clever if not always substantial jabs had left Huxley seriously bloodied. In a letter to a friend written a few days later, he noted, 'Had quite a long fight with Huxley. I think I thoroughly beat him'. This assessment was shared by the correspondent of the *Evening Star*. He remarked on the 'great power and eloquence' of Wilberforce's address. Twenty years later, the Bishop's son mentioned the Oxford debate in passing in his father's biography, and noted that his 'eloquent speech' against Darwinism had 'made a great impression'. The London daily *John Bull* agreed. Wilberforce, its columnist noted, had shown Darwin's *Origin of Species* to be 'built on very slight foundations'. Perhaps most astonishingly, it is far from certain that Wilberforce ever raised the genuinely delicate issue of Huxley's pedigree. Even friends of Huxley thought he might have taken offence where none was intended. If Huxley had landed a killer blow, it may have been on a shadow.

The recollections of the Darwinian botanist Joseph Hooker also diverged markedly from the received account (no prizes for guessing the identify of the hero in his version). Hooker told Darwin he had 'smacked' Wilberforce amidst hearty applause. At the end of his tirade, he boasted, 'Sam was shut up',

> [He] had not one word to say in reply & the meeting was *dissolved forthwith* leaving you the master of the field after 4 hours of the battle . . . I have been congratulated & thanked by the blackest coats & whitest stocks in Oxford.

Hooker's testimony is especially interesting for the light it sheds on Huxley's famous riposte. Huxley felt himself to have been 'the most popular man in Oxford for full four & twenty hours afterwards', but, according to Hooker, 'he could not throw his voice over so large an assembly, nor command the audience; & he did not allude to Sam's weak points nor put the matter in a form or way that carried the audience'. Far from delivering a beautifully apposite thrust against which there was no reply, several otherwise favourable witnesses recalled that Huxley was 'white with anger' and too excited to 'speak effectively'. A sheer excess of emotion had paralysed his tongue.

Indeed, the possibility is raised that the triumphalist account Huxley sent to Dr Dyster was more a face-saving device than the proud reflections of a man re-enacting the moment of his ultimate victory. This also accords with accounts of Huxley's temper at dinner the night of the debate. Eyewitness reports do not provide the image of a man basking in glory. And, like many others, Darwin—too physically and psychologically fragile to attend the meeting in person—was left in a state of confusion as to who had actually won the contest. The opinion of *The Athenaeum* magazine probably came closest to providing a fair assessment. The protagonists, it summarized, 'have each found foemen worthy of their steel, and made their charges and counter-charges very much to their own satisfaction and the delight of their respective friends'. This was not a night, however, of clear-cut winners and losers.

So the popular account sketched at the beginning of this essay is wrong in significant points of detail. The myth of Wilberforce's humiliation was constructed after the event by evangelistic Darwinians who relied on Huxley's carefully crafted recollections and turned a blind eye to counter-claims. In so doing, Huxley's supporters deliberately implanted in the minds of those who followed them the idea of a natural opposition between what Francis Bacon termed the 'works' and the 'words' of God. The legend they formulated presented Oxford 1860 as the point at which science finally broke free from the chains of dogma and turned its back on centuries of enforced silence on the crucial question of man's place in nature. Yet this image, too, is almost entirely bogus.

Science and religion in harmony

That there existed a fundamental distinction between science and religion would have been a hard concept for much of the audience of June 1860 to swallow. To most of those attending the debate, science and religion were fully reconcilable. To many, they would have seemed ultimately indistinguishable. Thus, when the dons and scientists trooped into the Oxford Museum they entered what could be described without jarring irony as a 'temple of science'. An ornately carved angel looked down on them as they walked into a museum, the construction of which Bishop Wilberforce himself had enthusiastically promoted. Nor was this the limit of his involvement in scientific affairs. Wilberforce was also vice-president of the British Association for the Advancement of Science and had himself delivered a scientific paper on the day before the debate. Self-evidently, 'Soapy' Sam was no enemy of science. Nor was he a humourless, inflexible ideologue. To him are credited the following irreverent lines of doggerel:

> If were a cassowary
> On the plains of Timbuctoo,
> I would eat a missionary,
> Cassock, band, and hymn-book too.

If nothing else, this short poem reveals a marked capacity to look at the world through the eyes of others.

Wilberforce's outlook on science was far from uncommon. Even late into the nineteenth century numerous scientists saw little in nature that was inconsistent with what they read in the Bible; and after minor concessions had been made to palaeontological and geological discoveries, many were satisfied with the general—if not always literal—veracity of Genesis. Indeed, dozens of mid-Victorian scientists claimed that science taught a new reverence and awe for the Almighty Creator. 'Scientific enquiry', announced Lord Wrottesley at the opening of the 30 June debate, allows man to 'come nearer to God'. One of the nineteenth century's most productive areas of zoology and botany—natural theology—was actually founded on the idea that by showing the complexity of Design in nature it became churlish to deny the existence of a Creator-God. This movement reached its peak during the 1830s with the publication of the famous *Bridgewater Treatises*, commissioned by an ageing Earl of Bridgewater as a means of expiating the sins of a dissolute life. The book's authors assembled what seemed to them overwhelming evidence that where science and religion are not competing for the same ground, a mutually beneficial exchange of ideas is entirely possible.

This is not to deny that there were signs of conflict to come: materialism—a belief in the purely physical nature of mind—was on the rise. So was evolutionism. As we saw when looking at Darwin's ideas (Chapter 9), in 1844 the Scot Robert Chambers had anonymously published *Vestiges of the Natural History of Creation,* a book which, amongst other things, seditiously claimed that man had evolved from rudimentary organisms. By the time Darwin's *Origin of Species* was published in 1859, however, the scientific and religious establishment generally felt that these threats had been contained by a judicious combination of accommodation and reasoned refutation.

Principally because it was written by a respected gentleman scientist and not a middle-class radical, *The Origin* was recognized to be a more serious challenge. Even so, standing before a scientific audience in Oxford in 1860, Bishop Wilberforce did not feel that science and theology would now have to part company. With the backing of the majority of the assembled scientists, and primed by Richard Owen, Britain's most-renowned palaeontologist at the time, Wilberforce subjected Darwinism to a series of cogent scientific assaults. From Wilberforce's own published review of *The Origin*, we can make a good guess as to what he would have said.

Show me a single historical example, Wilberforce presumably

demanded, of a non-domesticated animal species changing in structure over time. Mummified animals found in the pyramids of Egypt, he probably declared, were anatomically no different from their modern-day counterparts. Why had natural selection done nothing to change them over the past four millennia? In all likelihood, this was followed by an equally telling demand that Darwinists provide a solid piece of fossil evidence showing a species change into another over time. These would have been no baseless rhetorical flourishes; they struck at the heart of Darwin's argument. Lacking a compelling theory of inheritance and with only a scattering of fossils, natural selection remained vulnerable to such critiques well into the twentieth century.

Clearly Wilberforce had no particular need to invoke God. Bible-thumping was a defensive manoeuvre that he had no use for. Instead, he rapidly showed his ability to attack the opposition on their own terms and on their own ground by impugning the empirical tenets of *The Origin*. Accordingly, when this was reported to poor ailing Charles Darwin he was obliged to concede that Wilberforce's objections to the theory of evolution by natural selection were decidedly 'clever'. No wonder, then, that most listeners felt that 'honours even' best summed up the outcome of the debate.

The Young Turks

Given such an inconclusive result, why didn't the Darwinist just let the whole episode fade away into obscurity? An important clue to understanding why the event has assumed such symbolic importance lies in the fact that, as several members of the audience commented, 'The younger men were on the side of Darwin, the older men against him'. Darwin also observed that the success of his theory would depend on 'young men growing up and replacing the old workers'. This generational divide had little to do with proverbial old dogs and tricks. Instead, Huxley and his allies had a strong vested interest in asserting a sharp distinction between science and religion. Almost literally the vanguard of an army embarked on invasion, they needed a justification for war. To understand why, even if Wilberforce didn't, we do need to look into Huxley's ancestry and consider the social circumstances that separated him from the Bishop.

Despite great zeal and ability, Huxley had experienced enormous difficulty in breaking into the world of science. This is partly because he was not born a 'gentleman', a term integral to the way in which Victorians saw themselves and others. Although no one could precisely define what a 'gentleman' was, everyone understood perfectly what it meant. A key requirement, all agreed, was to have been well-born. So, as the son of an underpaid schoolmaster, Huxley was hardly even a contender. Income was another important consideration. With 'self-made men' in general being looked down on, Huxley's apprenticeship as a surgeon among the 'troglodyte' poor of London's East End earned him no credit. Yet another part of the definition had to do with the morals of one's family. Again, with one sister having to flee the country in suspicious circumstances, and another invariably drunk on gin or laudanum, Huxley had almost no chance of earning gentlemanly status. His saving graces were all personal. On integrity, temperance, and forbearance he scored highly. He also had the bonus of being markedly intelligent and superbly witty. Nonetheless, it was only after spending several years as a surgeon on HMS *Rattlesnake*, and deeply impressing an aristocratic sponsor, that he secured both his entrée into the world of professional science and grudging acceptance as a gentleman.

Yet, still, after gallons of perspiration and years of privation, during the 1860s Huxley was struggling to make ends meet. He may have been a world-famous scientist, and had at last been accepted as a bona fide gentlemen, but he could hardly afford to pay the grocer. Huxley's problem was that Victorian science was still largely the province of the skilled amateur. In an age of privilege and snobbery in which few scientists received an income for their research, science did not yet exist as a distinct profession. To receive pay for pursuing scientific interests would have been considered undignified, if not insulting. And rather disingenuously given their personal wealth, well-bred men of science argued that the pursuit of pure truth should not be sullied by the mercenary craving for cash. In 1865, when Richard Owen, another lowborn biologist, told a countess that he was a salaried scientist she actually recoiled in horror.

To modern minds at least, the underlying objective was to enable wealthy gentlemen-scientists, and professionals able to fund their scientific research with the income from another calling, to keep out the 'lower orders'. Certainly these amateur scientists, many of them clergy-men, had little interest in establishing the paid university professorships on which depended the hopes of those—such as Thomas Huxley—born into lower social strata. As a result, in 1860 there was no more than a handful of state-funded centres of scientific research within the United Kingdom. So, having clawed his way into a respectable position, Huxley rapidly grasped the awful fact that, unless things changed, he would spend a life kowtowing to wigged, dog-collared, and mitred amateurs without ever earning enough to keep the bailiffs from his doorstep.

The solution for this dynamic and highly intelligent young scientist was simple but radical. He knew that the country was valuing science more and more. Even the notoriously stingy civil service was worrying that the British lead in science and industry was being challenged by the Germans and the French. Science clearly had a bright future. In this context, Huxley decided that the state would not hesitate to pay decent wages to the salaried scientist if the majority of amateurs could be turfed out or reduced to the status of mere collectors.

But how to topple the existing elite? This is where the Oxford debate ties in. Huxley also saw that most amateurs were rather orthodox in their Christian beliefs. Most of them could hardly afford to be otherwise. Regular church attendance was a mark of the sober gentleman. So great was the pressure to conform, that some free-thinkers joined two churches so that the congregations of each might give them the benefit of the doubt when they stayed at home on Sunday mornings. In this situation, Huxley saw a marvellous opportunity to force out the wealthy amateurs by driving a wedge between their science and their religion. If they could be made to feel that they could not pursue a scientific calling without sacrificing their most basic religious principles, their code would require them to forego the former. The resultant exodus of amateurs from the higher ranks of scientific debate would permit Huxley's coterie of scient-ific careerists to assume the reins of power. Then they would gain the financial and the social rewards they felt to be their just deserts. It was for this reason that, as Huxley wrote to Darwin in 1859, he was 'sharpening up [his] claws and beak in readiness' for a fight.

Enemies of science

From the late 1850s onwards, Huxley set about 'unmasking' Establishment science as intellectually bankrupt. His mantra 'Science versus Parsonism' encapsulated both the new model of science he was advancing and the means by which this 'young guard' would pull off their coup against the traditional scientific elite. Deliberately redefining science in terms inimical to Establishment religion, Huxley even coined the term 'agnostic' to describe the only attitude he believed a legitimate scientist could adopt towards religion. Thus, within a few years the radical wing of science had become agnostic, thoroughly anti-scriptural, and a very powerful lever for ousting the amateur from the scientific domain. Happily invoking the fate of Galileo at the hands of the Catholic Church, Huxley warned the public of its duty to 'cherish' science and defend it against 'those who would silence and crush her'. In Huxley's rhetoric, those with dual loyalties to science and religion were portrayed as representing forces corrosive to the advancement of science, smothering her spirit of free enquiry. And it is in this context that we need to set the Oxford debate.

To those supporting Darwin's ideas, the answers given by evolutionary theory to perennial questions in natural history were far less important than the platform they provided from which to discredit scientific amateurs. By striving to make their radical views the new orthodoxy, the Young Turks were creating a situation in which, no matter how competent his research, the amateur could be told without ceremony that he was no longer welcome at such high tables of science as the Royal Society.

Unsurprisingly the polarized image of science and religion that Huxley presented was considered strange and alien by many practising scientists. The young guard was fighting a boundary dispute where few of their opponents had realized there was a boundary to be contested. But relentless attacks on 'interfering' prelates and pious 'meddlers' soon got the message across. The well-connected, well-heeled, and impressively articulate bishop who faced Huxley and Hooker on that June evening in 1860 embodied for these reformers the Establishment power that stood between them and professional success. But Wilberforce also exemplified another characteristic of the amateur which Huxley & Co. sought ruthlessly to exploit. Like many amateur scientists, his knowledge was broad but shallow. This meant that he could be unfairly portrayed as an arrogant dilettante, cheapening science with his ill-informed sallies into complex debates. 'I showed Wilberforce to be absolutely ignorant of the rudiments of Bot[anical] Science', Hooker proclaimed in the aftermath of the Oxford debate. And the *Evening Star* correspondent recalled that Huxley had branded 'his Lordship' an 'unscientific authority'.

Where once the scientific community had esteemed the generalist, the new man of science was to be a specialist, entirely unashamed of his ignorance of anything beyond his immediate intellectual frontiers. This new reverence for specialization—still seen by many as essential to scientific advance—first came into being as a stick with which to chase the amateurs out of the yard. Moreover, to the decision by this cabal of reformers to redefine science may be traced much of the rhetoric of the scientific method. Claiming a commitment to the rational apprehension of the truth without ulterior motive and somehow hermetically sealed off from wider influences, these brilliant men changed the way people thought about science and the scientist. The ideal they formulated was deliberately congenial to bright young men who could benefit from the introduction of a salaried career structure. It boded ill, however, for the amateur and the polymath. Over the following years, science became a less and less popular pastime for the cleric and the well-to-do amateur. And, as the field became almost exclusively the preserve of the university and the laboratory, the new men rewrote history to provide the quintessential foundation myth. Hence the yawning gap between what is thought to have happened and what actually happened on the night of 30 June 1860.

The aftermath: conflict or compromise?

Having the liberty to write their own history, the evangelical Darwinians were also free to impose their polarized view of science and religion on accounts of the aftermath of the 1860 debate. Dozens of histories and biographies have been written on the assumption that Darwin's *Origin of*

Species annulled irrevocably the abusive marriage between science and religion, leaving the church to pine for its loss. But once again this standard account is largely mythical. Mid-Victorian religion was no mono-lith and its portrayal as an inherently reactionary force is plain wrong. As early as the 1840s, a liberal wing of the Anglican Church—paralleling the reformism of Huxley's clique within science—had been asserting that much of the Bible should be treated as a historical text devoid of spiritual significance. These revisionist clerics triggered a series of internecine conflicts among the clergy on a scale that far exceeded the inconclusive jousting between Huxley and Wilberforce.

Huxley & Co. even exploited the Anglican Church's internal conflicts to their own advantage. They showed no scruples when the opportunity came to ally themselves with a dissident but influential group of churchmen. And in 1860 they signed a letter protesting at an attempt by Wilberforce, the Archbishop of Canterbury, and 25 bishops to have several liberal theologians arraigned before the ecclesiastical courts. In subsequent years, they forged strong links with several senior clergymen and it was one of these alliances that enabled them to pull off the remark-able coup of having Charles Darwin buried in Westminster Abbey in 1882. The Revd Frederic Farrar, Canon of Westminster and a close friend of Thomas Huxley, happily provided this notorious infidel with full Christian rites. Self-evidently, progressive churchmen and reformist scientists could still make common cause even after the Oxford 1860 showdown.

The view that a complete estrangement of science and religion followed directly from the 1860 debate has some other uncomfortable facts to overcome. First, well into the twentieth century, only a few scientists accepted the atheistic overtones of Darwinism. Second, a large proportion of late-Victorian clergymen made substantial concessions to evolutionary theory.

Let's take the scientists first. Charles Lyell, the nineteenth-century's greatest geologist and Darwin's firm friend, was profoundly impressed by the evolutionary arguments of *The Origin*. But he was never prepared to 'go the whole orang' and accept that God had played no part in the appearance of human beings. In his 1863 book the *Antiquity of Man*, this internationally renowned scientist presented an emasculated version of

Darwinism, in which Divine intervention had been necessary to raise man above the level of his bestial progenitors. This gravely pained Darwin, but worse was to come. During the late 1860s, Alfred Russel Wallace, the co-discoverer of the theory of evolution by natural selection, became fascinated with spiritualism. Under its influence, he began to argue that a Divine power interceded at some stage during human evolution and bestowed on man mental abilities that could not be explained in terms of natural selection.

Meanwhile, across the Atlantic, Darwin's most distinguished supporter, the deeply religious botanist Asa Gray, sought to reconcile evolution and Christianity by arguing that God had inaugurated the laws of evolution and personally ensured that new variations regularly appeared. Darwin, Hooker, and Huxley cried 'apostasy!', but the purity of their view placed them, rather than Gray, in the minority. Nor was the position of the majority unreasonable. As we have seen, Wilberforce had no trouble in pointing out that the evidence in favour of non-theistic Darwinism was not overwhelmingly strong.

Turning briefly to the late-Victorian clerics, most of the upper echelons were soon happy to accept ideas of evolution in an attenuated form. They did this, like Lyell, Gray, and Wallace, by writing God back into the story as both designer and overseer. Nor were they alone in feeling that Darwin's ideas needed something extra. Perhaps the most striking feature of the post-1859 period is that scientists, churchmen, and the educated population at large became increasingly sceptical of the theory of natural selection. As seen in Chapter 9, the idea of evolutionary change driven by blind selective forces was unappealing to a religious society buoyed up with national optimism. In its place, most people preferred an evolutionary model that suggested either a built-in and divinely ordained tendency towards further improvement or an ever-watchful God who supervised the process of ongoing development. These ideas of inevitable progress suited the time and the place: the unregulated roulette wheel implicit in Darwinism—although not fully accepted by Darwin—most certainly did not.

As a result, by the 1880s, both scientists and non-scientists were strongly drawn towards the much more congenial religio-progressive worldview. Save for a few intractable conservatives, the Anglican

Church had by then joined the laity in accepting evolutionary theory as a reassuring indication of continuous progress and divine omniscience. Rather than having been forever put asunder, after 1860 science and religion continued to enjoy the most harmonious of unions. In fact, if we look in this late-Victorian English landscape for any group that did suffer from the rejection and reduced circumstances Darwinian legend now wrongly ascribes to the Established Church, one very obvious candidate would be those still clinging on to the theory of natural selection. For them, salvation—and the rewriting of history—was still beyond the horizon.

Post mortem

The traditional account of the Oxford clash between Thomas Huxley and Samuel Wilberforce is a serious distortion of what the records say happened. The night itself did not see a clear victory for the professional scientist. When the dust settled, religion in England was as dynamic and powerful a social force as ever. And Bishop Wilberforce was just as comfortable in his religious beliefs as he had been when a fresh and ambitious undergraduate. For a combination of empirical and cultural reasons, science and religion would continue to walk side by side for many years to come. Wilberforce himself would live a further 10 years before being killed in a fall from his horse (Huxley joked that on the one occasion that the Bishop's brain and reality came into contact, the result had been fatal). And although it is possible that during the Bishop's last decade the merits of Darwin's case were increasingly brought home to him, on balance, I think this very unlikely. Even among his less-conservative brethren, Darwinism remained a minority view, and most senior clerics found that a belief in a divinely ordained evolution of species was perfectly compatible with their religious sensibilities. So, in terms of the changing of minds, the Oxford debate was largely a non-event.

But although the exchange of angry words seemed to lead nowhere, at a deeper level profoundly important changes were being initiated. The real importance of the event lies in the fact that the tactics then inconclusively used by Huxley and his friends Joseph Hooker and John

Lubbock became their standard means of dealing with the clerics and gentlemen scientists who stood in their way. In Oxford, the Young Turks gave their strategy for usurping the amateur its first serious trial. Like the use of tanks at Cambrai in 1917, the first outing yielded much more evidence of potential than actual success. Thereafter, they would refine their approach, widen their attack, and eventually sweep on to victory. Before long, Huxley & Co. had internalized their own propaganda and were to campaign under the same legend for the rest of their lives: 'Science and religion are incompatible'. The amateurs were increasingly made to feel that they were aliens in what had been their own territory. Within a generation, they had given up what remained of their hold. And by the close of the century there was only one ordained clergyman at the Royal Society. A mere two decades earlier he could have met a sizeable proportion of the General Synod at every meeting.

This is the real import of the Oxford clash. It marks the point at which professional scientists started to apply the exclusionary principle with a vengeance. For the parson naturalist, although he may not have realized it until many years later, June 1860 marked the beginning of the end. A wedge was to be forcibly driven between David Hume's duality of 'faith' and 'reason'. And there, with only a few exceptions, it was to remain. Science was to be moved much closer to the levers of power, religion away from them. In Britain, unlike some parts of North America, professional science's victory would be so complete that counter-attack became unthinkable. Public opinion would be so shaped that there was to be no UK equivalent of the Scopes trial. Similarly anyone seeking to have Creationism treated as a viable scientific theory would be marginalized.

One hundred and forty years later, at Oxford's 'Science versus Religion' debates, almost of necessity, the religious belligerents were mostly drawn from the United States. Given as easy a brief, they might have performed just as well as Bishop Wilberforce. Unfortunately for them, the evidence in favour of evolution is now too good for them to have stood a serious chance of winning the day. To one tender-hearted listener at least, the overwhelming weight of the scientific arguments deployed against them induced an atmosphere kindred to that of a ritual humiliation. Rather than seeming chevaliers of the Truth, Richard

Dawkins and his supporters sometimes looked like unconscionable bullies. If nothing else, this testifies to the enormous distance professional science has come—both scientifically and ideologically—since Huxley's campaign began back in the 1860s.

PAINTING YOURSELF INTO A CORNER

Charles Best and the discovery of insulin

Charles Herbert Best physiologist who, with Sir Frederick Banting, was the first to obtain (1921) a pancreatic extract of insulin in a form that controlled diabetes in dogs. The successful use of insulin in treating human patients followed. But because Best did not receive his medical degree until 1925, he did not share the Nobel Prize for Physiology or Medicine awarded to Banting and J. J. R. Macleod in 1923 for their role in the work.

'Charles Best', *Encyclopaedia Britannica* (1992).

In 1922, the world witnessed a medical breakthrough that was as dramatic as it was profound. Comatose diabetics and diabetics on the verge of starvation were injected with a newly isolated compound called insulin. Administered by a team from the University of Toronto, comprising Frederick Banting, Charles Best, James Collip, and John J. R. Macleod, its effect seemed miraculous. Within days, patients rose from their beds and began to resume relatively normal lives, their whole existence transformed by the fruits of medical science; it was one of those breakthroughs that reminds us how fortunate we are to be living in a scientific age.

The story of insulin, a hormone that is still the lifeline of millions of diabetics all over the world, shows science at its very best. The vital discovery was based on the combined efforts of many generations of scientists from across the world. Doctors in Ancient Egypt had recognized that some people are unable to process sweet foods and therefore produce urine with a very high sugar content. Early nineteenth-century pathologists noticed a strong correlation between diabetes and a shrunken pancreas. Paul Langerhans, a German doctor, first identified the islets in the pancreas that would later be shown to manufacture insulin. Then

Left: Charles Herbert Best (1899–1978).

research teams in Germany, the United States of America, Britain, France, Romania, and Canada undertook extensive programmes of animal experiments to isolate the active compound missing from the pancreases of diabetics and develop it for clinical use. The ultimate success of Frederick Banting's Canadian team can properly be viewed as the pinnacle of a major international effort.

But this shining scientific triumph has a murky underbelly that highlights the less-pristine side of science. It throws into bold relief the vicious personal competition, conflict, and grudges that innervate many important scientific debates. Credit in science usually goes to those who cross the finishing line first. This often means that those lucky enough to be holding the baton in the final leg get almost all the plaudits whilst those who carried it during the earlier stages are denied a place on the podium. Given the numbers limitation, Nobel Prizes can throw up extreme examples of this. Bitterness frequently ensues when those who did vital preparatory work are ignored. In the case of the discovery of insulin, the history books often mislead by implying that insulin research begin in 1921 and ended in 1922, with only the Banting team being significantly involved. Not surprisingly, given the worldwide effort then being made, several teams felt that their work had been unfairly marginalized by the Prize Committee. Indeed, the overlooking of some parallel successes in Eastern Europe does raise some very serious questions about the way in which prizes are awarded. In this chapter, however, I focus on a much more protracted, bitter, and unreasonable feud involving one of the men involved in the 'winning' team: Charles Best.

We have seen how Joseph Lister worked hard to rewrite history in his own favour (Chapter 8). With Louis Pasteur, John Snow, and Alexander Fleming (Chapter 1, 6, and 12), others took a considerable hand in the process of image-burnishing; Lister worked assiduously on his own behalf. Yet, Lister's considerable manipulations of the truth are overshadowed by the distortions effected by the Canadian scientist Charles Best. For Best not only rewrote the past but he pursued those who he felt had stolen his glory with a malignity that borders on a vendetta. His almost pathological craving for intellectual recognition drove him to discredit ex-colleagues with such passion, conviction, and consistency that today his version of events is still widely accepted in the medical community. Two of the many elements of the myth he created are included in the quotation at the head of this chapter: first, that only a technicality stood between him and a Nobel Prize; and, second, that his 1921 work on dogs unequivocally established the role of insulin in combating diabetes.

Recent research has shown there to be a yawning gulf between what Best claimed and what actually happened. It is crucial, therefore, that we first get the facts straight. Mostly thanks to the exhaustive labours of a single historian of science we can now outline these with impressive certitude. Michael Bliss's research allows us to understand what really happened in the build up to the Canadian success, and in the years that followed, to a degree that has never before been possible. The following account is almost entirely indebted to his groundbreaking work.

The facts of the case

As has been made clear, by the 1920s teams in various countries were working on the relationship between the pancreas and diabetes. Much of this work was vivisectionist, entailing the removal of the pancreases from dogs to induce the disease. Without a pancreas the dogs could no longer either burn off sugar or convert it into fat. In every case they rapidly slipped into comas and eventually died. Researchers learned from these experiments that the pancreas produces more than one chemical and they rightly concluded that the one lacking in diabetes patients was manufactured in the area of the pancreas known as the Langerhans ducts. The challenge lay in extracting a pure form of this chemical to see if it could protect a dog without a pancreas from premature death. Being able to perform this key experiment required several major technical advances. First, the active chemical had to be identified. Second, it would then have to be separated from the other chemicals produced by the pancreas. Third, a way would have to be found of administering it in the gradual way the body needs.

The work in the University of Toronto laboratory began soon after Frederick Banting, then a little-known general surgeon, read a scientific paper on the effects of blockages in the pancreatic ducts. Decorated for bravery in the First World War, Banting had struggled to make his mark as a medical researcher in its aftermath. This was soon to change. In the evening after reading the pancreas paper, Banting jotted the following triplet in his notebook:

> Diabetus.
> Ligate pancreatic ducts of dog. Keep dogs alive till acini degenerate leaving Islets.
> Try to isolate the internal secretion of these to relieve glycosurea.

Banting proposed the use of ligatures because, in addition to insulin, the pancreas produces enzymes (from the acini cells) that break down the secretions of the Langerhans ducts. In normal circumstances this is not a problem. But it becomes one when researchers start interfering with the body's finely tuned system. Previous attempts to make progress by removing the pancreas and then injecting material extracted from it into experimental animals had all failed because the insulin had already been destroyed by the pancreas itself. The paper Banting read suggested that ligating might offer a way round this problem. If the pancreas were sealed off from the rest of body and allowed to wither away, the islets of Langerhans would be the last to atrophy. This presented the possibility of isolating their contents so that they could be ground up alone and administered to an animal that had had its pancreas removed. If the ducts did contain the anti-diabetic compound and therefore ameliorated the animal's condition, Banting knew that he would be a long way down the road to a new therapy of exceptional importance.

This was not a novel approach. It had already been tried by an American team several years earlier. They had obtained some positive results but when the team leader's career took him elsewhere, the project petered out. Banting, however, had no knowledge of this, and a few days after reading the article, made an appointment with the head of the Toronto physiology department, J. J. Macleod. After some prevaricating, Macleod gave him lab space, a few animals, and a research student. Two research students were at that time available, Charles Best and Clark Noble, and there is good evidence that they flipped a coin to see who would work on Banting's team: Best won the toss.

In March 1921, work began. Banting and Best's experimental design, on which Macleod advised, involved removing the pancreas from some dogs and ligating the pancreas in others. After a few weeks, and several failures, the ligated pancreases were removed, the Langerhans ducts cut

Frederick Banting (1891–1941) in his Toronto laboratory.

off and processed, and a small amount injected into the depancreatized dogs' veins. The results were far from clear cut. One dog was brought back from a coma by the extract, but died the following day. Still, several of the dogs did become more active after the injection and there were enough positive signs to give Banting and Best the encouragement they needed. Macleod had been on holiday in Scotland while the experiment was being carried out. On his return, though intrigued by some of their results, he was far from impressed by their technical skills. Monitoring each dog's blood-sugar level before and after each injection was crucial in this sort of experiment. If the blood-sugar level went down it could reasonably be claimed that the preparation contained a hormone that could control diabetes. But, far from being scrupulous in this, Macleod found that Banting and Best had performed these tests infrequently and unreliably. Macleod strongly recommended that they conduct future experiments with much greater precision and with many more controls.

In response, Banting and Best began to undertake proper pre-injection blood tests. To complement these, they also started to take more frequent blood samples after injecting the dogs with extract. Several months later, they published their first scientific article. In it the two men explained how they had 'always observed a distinct improvement in the clinical condition of diabetic dogs after administration of extract of degenerated [i.e. ligated] pancreas'. As we have grown to expect, this was not quite the case. It is clear from their notebooks, and the difficulty later scientists had in reproducing their results, that Banting and Best were highly selective in which experiments they included in the paper. Reminiscent of Louis Pasteur, Arthur Eddington, and Robert Millikan, successes were exaggerated and failures sometimes quietly ignored.

By this stage, despite extensive experimentation, Banting and Best had made little progress beyond that achieved by other groups of whose existence they were only slowly becoming aware. Then, joined by a visiting biochemist called James Collip, they started to make real advances in administering subcutaneous injections and in the accuracy with which they were able to test the dogs' blood-sugar levels. On the suggestion of Macleod, the extraction of what became known as insulin was also improved with the use of alcohol to separate it out from the other residual chemicals in the pancreas. Most importantly, the team now found that they could experiment with fresh, adult pancreases so long as they used appropriate filtrates. Crucially this meant that tricky ligation operations were no longer required. In all of these areas, the expertise of Macleod and Collip was vital. Collip's contributions were especially fruitful. On his own initiative, he performed numerous experiments with rabbits in which extract lowered the blood-sugar level of even healthy animals. Then he made a slight change to the extraction process that would pay a major dividend. After the final filtration stage he decided to keep the residue as well as the filtrate and then compare their effectiveness. Contrary to expectations, it soon became clear that, of the two, the residue was much the more effective.

Even before this stage was reached, Banting and Best had begun to feel that Collip—an experienced and very effective scientist—was beginning to assume control of the project. Now his successes were starting seriously to disturb the original pairing. They almost certainly had grounds for this. Whereas Collip had the gifts of an outstanding researcher, the evidence Michael Bliss has built up suggests that Banting and Best were appreciably less able. Perhaps in a bid for triumph while they were still in charge, they decided prematurely to test their pancreas extract on a human diabetic. On 11 January 1922, the first clinical trial was performed on a young man called Leonard Thompson. Results were very disappointing. Only slight benefits, which could easily have been due to other factors, were discerned. Moreover, Thompson acquired a large and unsightly abscess at the site of the injection. Banting and Best had overreached themselves and they knew it.

In the aftermath of this test, Banting was honest enough to recognize that this experiment had not demonstrated the efficacy of his pancreas extract. His and Collip's group then signed an agreement to work together rather than trying to compete: neither side wanted to risk being pipped at the post. And given the outcome of the foolish risk they had run, Banting and Best were now ready to admit that they actually needed Collip's help. Later in the affair, this is something that they would conveniently forget.

During the period leading up to the Leonard Thompson trial, Collip had been concentrating his efforts on extracting and purifying the active principle. The technical refinements he had introduced were now producing a much more potent extract. Just weeks after Banting and Best's failure, this new extract was used on humans. The results were profoundly gratifying. As Michael Bliss has put it, 'extract purified by Collip brought dying diabetic children back to life and health'. It was a truly momentous occasion and those involved must have recognized it as almost certainly the greatest moment in their scientific careers. Within two weeks, on 3 May 1923, the discovery of insulin was announced by Macleod with befitting fanfare to the medical world. A year later, the Nobel Prize in Physiology or Medicine was shared by Frederick Banting and J. J. Macleod. According to several later polls, Fred Banting became the most famous Canadian in history. Even today he remains one of his nation's most cherished sons.

From colleagues to rivals

This was one of those cases in which it can fairly be said that the awarding of the Nobel Prize caused more trouble than it was worth. Macleod's acceptance speech may not have helped matters by going to exceptional pains in putting Banting and Best's contribution into historical and international perspective. At great length Macleod pointed out just how much had been done by other teams in other countries. He also made clear that it was James Collip who produced the high-quality extract. In the aftermath of the award ceremony, Banting found a means of indicating his belief that the prize should have been awarded to himself and Charles Best, by sharing his prize money with his junior colleague. Macleod responded to this by giving half his award to Collip.

The Nobel Prize was not solely to blame for the insulin team's troubles. It would seem that relations within the Toronto insulin team had rarely been cordial. Before the contretemps over the Leonard Thompson affair, Best and Banting themselves had almost come to blows when the latter had criticized his assistant's research methods. The healing of this rift seems to have been brought about by both recognizing the greater threat presented by Macleod and Collip. Henceforth their animosities were turned in that direction. These feelings were merely intensified by the Nobel award.

There was more than just sympathy for Best in Banting's annoyance at Macleod's prize. Because Macleod was the Head of the Department, Banting became concerned that others might cast him as Charles Best, the mere assistant, to Macleod's Banting. He would have felt much more secure had the Nobel award made explicit the fact that Charles Best played Charles Best to Frederick Banting. This notwithstanding, he was initially faced with a straight choice between letting things lie or publicly repudiating Macleod's importance. To begin with, he chose the latter.

All nations love their heroes, so, understandably, Canadians coveted the story of insulin's discovery. In this lay Banting's sterling opportunity to alter the record. With each of many retellings of the story, he gradually wrote Macleod and Collip out of history. Directly and by implication, Banting got over the message that almost all the credit was due to him as the one whose crucial insight was contained in his notebook triplet. This in itself was misleading. Nothing in those three lines added to the contents of the article he had read and to what had already been tried elsewhere. Even more to the point, ligating the pancreas was not the method that the team used a year later to provide extract for the successful treatment of patients.

Banting also achieved his ends by emphasizing the quality of the results he had obtained when working without the help of Macleod and Collip. Yet, as we have seen, not only did these two men play crucial parts in the development of the successful treatment, but, working alone, Banting and Best met with more failure than success. Indicatively the only clinical trial performed without the direct involvement of Collip had been, at best, clinically inconclusive. It is the considered opinion of Michael Bliss that 'Banting's and Best's research was so badly done that, without the help of Macleod and Collip . . . the two young Canadians would be fated to disappear from medical history'. Nonetheless, after a decade or so, Canadians were being taught that insulin had been discovered by Banting and Best alone.

Nationalism may also have played a part in the ease with which Macleod's contribution was eclipsed. He, after all, was a Scot who returned to Scotland. But although Collip, like Banting and Best, was a Canadian, he presented too big a reputational threat. Banting did not want to share the glory with a man who soon gave further demonstrations of his outstanding professional abilities. If there was to be any sharing, it would be with Best. To Banting, Best's great advantage was that he could be firmly characterized as the assistant. So he never willingly credited Best with anything more than offering practical help. And even in Banting's most generous of moods, the younger man was described only as a source of emotional and practical support. The senior man never attributed any particular insights or discoveries to his junior. The implication was always that had the spin of the coin made Best's colleague, rather than Best, Banting's assistant, history would not have been materially different.

Best comes in from the cold

Given this self-serving diminution of Best's role, it comes as no surprise that relations between the two men deteriorated in the aftermath of Banting's seemingly noble act of sharing his prize money. As Best embarked on a lifetime's campaign to ensure that he got what he considered his due share of the glory, he increasingly alienated Banting. By the end of the 1930s, the older man's feelings towards his onetime colleague bordered on loathing. Michael Bliss suggests that Banting may, in any event, have been uneasy with the nature of his triumph. Certainly the ferocity of his temper was often remarked upon. But there could be no doubt that his animosity towards Best was exceptional.

Perhaps encouraged by the fair degree of research success he enjoyed after 1923, Charles Best wanted a research institute to be established and named after him. As this honour had already been conferred on Banting in 1930, it was clear that Best was seeking parity of esteem. Incensed at what he considered alarming effrontery, Banting immediately set about frustrating these plans. 'Best is naive in his abject selfishness', he fumed. Then, in 1940, no doubt in part to heighten his profile, Best offered himself as the Canadian government's medical emissary to a besieged Britain. He lobbied hard but as soon as he got his way he pulled out. Banting's contempt was such that, fully aware of the dangers involved in the crossing, he elected to go himself. Shortly before leaving he was heard to say, 'If they ever give that chair of mine to that son of a bitch, Best, I'll roll over in my grave'.

Tragically Banting's plane crashed in Newfoundland in early 1941 killing all on board. The event robbed Banting of a chance to clear his conscience with regard to Collip. One of his last wishes before leaving Canada had been that his calumnies against his erstwhile colleague be erased and Collip's role properly recognized with an honorary degree. This plan was quickly scotched when Best was appointed Banting's successor. Banting's corpse must have been spinning like an electric turbine as, in the following weeks and months, Best was given both his chair and control of his department. Banting's old administrator immediately resigned in disgust, writing to Collip soon after, 'I went so far as to tell the President [of the University] that I felt it was the last person Dr Banting would want to succeed him in the department'. To no avail. Wanting a big gun it was easy for the President to suborn consideration of a dead man's wishes, no matter how great he had been. The discovery of insulin had bought Toronto a place in the sun. Having Best replace Banting was the most effective way they could see of extending the lease.

In 1935, John J. R. Macleod had also died. This meant that for Best the way was now as clear as it was ever likely to be for him to rewrite the insulin story, assigning to himself the lead role. The plan he embarked on required an extraordinary combination of gall and self-delusion. It also reflected an almost pathological need for intellectual approbation. His first major opportunity came in 1946 when the twenty-fifth anniversary of the discovery of insulin was being celebrated.

Bear in mind, first, that Banting had never spoken of Best as an equal partner, even before he began to despise him; and, second, that this interpretation is supported by their laboratory records. In 1946, however, Best set about putting a very different account into the dead man's mouth. Speaking to the American Diabetes Association, he declared:

> As Banting has stated, very clearly we began work in partnership. Indeed that was the only possible relationship when both of us were without a stipend and each was responsible for a definite aspect of the research . . . it is difficult to imagine a closer working arrangement than that which developed between us.

These were finely crafted sentences. Similarly a meagre one-sentence reference to Macleod and Collip simultaneously paid them some due and implied that their contributions deserved no further elaboration. Such remarks were highly charged with the intense emotion of a man desperate for credit. But in their manufactured quality, they also betray Best's awareness of what he was doing.

In this pivotal speech, Best made other tendentious claims. First, that he had realized the use of alcohol would much improve filtration several weeks before Macleod intervened to suggest it. Second, that in their early experiments leading up to his and Banting's first published paper they were able to control 'all the signs and symptoms of diabetes in dogs'. Third, that he personally made the pancreas extract given to the first 'human subjects'. For good measure, he also added the groundless claim that both he and Banting had been drawn into diabetes research having seen people close to them fall victim to the disease. Over the next few years, this suite of claims became Best's mantra. The falsehoods and half-truths first aired in 1946 were reiterated time and time again. Yet, having had full access to the original papers, Michael Bliss argues that the truth of the matter is that everyone involved contributed vital ideas and expertise with the possible exception of the youngest and least experienced, Charles Best. So we may now briefly examine the validity of Best's 1946 claims.

We can be sure that alcohol filtration was undertaken only on the advice of Macleod (see below). Throughout 1921 and 1922 the flow of advice from Macleod was intermittent but vitally important to Banting and Best. He was familiar with much previous work on diabetes and knew the different methods of isolating compounds. Just as significantly, unlike Banting and Best, he knew what a well-executed experiment looked like. It was under his direct tutelage that these two men learned the importance of strict controls and careful observation. Only after Macleod's return from holiday in 1921 did their experiments start to become reliable.

It follows from this that Best's second claim was also false. Banting and Best's early work on depancreatized dogs was not conspicuously successful. Attempts to keep the animals alive for long periods of time with pancreas extract almost invariably failed. Moreover, because their experiments were so sloppy, there was no certainty that the results they had obtained had anything to do with the extract at all.

Third, the addition of Collip's expertise was essential for the carrying out of successful clinical trials. The extract they had used before his intervention was too impure and the reactions produced too ill understood for success to be at all likely. The all-important difference in level of success between the Leonard Thompson trial and those performed in February directly testify to Collip's contribution.

Fortunately for Best, most of the diabetologists around the world were not privy to these details. Far easier to accept the words of a revered scientist than to go back and look at scientific papers and notebooks that were by then more than 20 years old. As a result, by the 1950s, only close friends of Banting, Collip, and Macleod were left grumbling. In the minds of many, all over the world, almost everything was due to Banting and Best. Indeed, by the time Best had got fully into his stride, some may even have been led to question Banting's role.

In 1953, the seal seemed to have been set on this story by Best's close friend, the accomplished British physiologist Sir Henry Dale. The occasion was the opening of the Best Institute at the University of Toronto. Ignorant of the contributions of other researchers, Dale sincerely believed in his friend's version of events and felt that he had been the victim of a gross injustice. There is also evidence that Dale may have had an old professional score to settle with the ghost of Macleod and was glad of the opportunity to downgrade Macleod's role. But, for whatever reasons, Dale was happy to use his speech opening the institute to affirm Best's reconstruction:

> The collaboration was to be one of intimate understanding, with no question between the two participants of any but an equal sharing of its success . . . Macleod, still quite naturally sceptical of any successful outcome to the enterprise, left Toronto to spend the summer in Europe; so that it was in an otherwise deserted Department that the two young and inexperienced but determined enthusiasts . . . solved the main problem without further aid from, or communication with, anybody. As a result they had clear evidence of the existence of insulin, and of the possibility of obtaining it in a separate solution, and of eliciting its effects by artificial injection, by the time Macleod returned from Europe.

The audience at the University of Toronto, as might have been expected, clapped with great enthusiasm. But, just as Best seemed to be winning the recognition he craved, his plan began to unravel. The problem was that the language of his own and his supporters' remarks had lost the subtlety of his 1946 speech. Blatant falsehoods and egregious exaggerations were now being committed to the public record. Even the reticent and surprisingly modest James Collip began to feel his hackles rising.

The muddying of the waters

Collip was in the audience on the day Sir Henry Dale delivered his speech. And he was not best pleased. A few weeks later, Dale heard from a third party of Collip's profound irritation. Well primed by Best, Dale immediately assumed that his friend was once again being unfairly maligned. To be absolutely sure, he wrote to Best and asked for his reaction. Best's knee-jerk response was crude and defensive, 'This is not the first hallucination which Collip has had in recent years'. It was followed up by a seven-page defence of his account and of Collip's relative insignificance in the discovery. Most of his claims in this document lack any solid foundation. For example, Best said that before Collip had become involved he had purified enough insulin to keep a depancreatized dog alive for 70 days. He did not mention that an autopsy showed that the depancreatization of this dog had been only a partial success. Likewise, he skated over the fact that Leonard Thompson's first injection with pancreas extract had been of doubtful importance and had had negative side-effects. Although this confection seems to have satisfied Dale, it meant that Best had given yet more hostages to fortune.

Best soon encountered individuals markedly less willing to take his story at face value. In 1954, the National Film Board of Canada decided to make a film about the discovery of insulin, one of the jewels in the nation's crown. At first, Best and his admirer, William R. Feasby, thought this was an excellent opportunity to get the Best version of the story to a very wide audience indeed. When draft transcripts were sent to them, these men took foolhardy risks by picking and choosing which elements of the conventional story they wished to leave intact and which could be written up to emphasize Best's scientific genius and capacity for self-sacrifice. The tossing-a-coin incident was erased and replaced by Best's bogus recollection of forcing himself on Macleod and Banting after a close relative had died of diabetes. Best even invented entire sequences of dialogue in which he was seen lecturing Banting on scientific issues in the light of knowledge that became available only many months later. In short, Best's reminisces were extravagantly self-indulgent and self-serving. Unfortunately for him, however, he had not counted on the assiduity of the Film Board's writer, Leslie Macfarlane.

Macfarlane was a thorough man. Noting major discrepancies between Best's account and a summary of events prepared by Macleod shortly before his death, the film-maker went back to the original sources. These included unpublished accounts of 1921 and 1922 written by Banting and held by his widow, as well as contemporaneous laboratory records. Against the latter there was no arguing. What Macfarlane found forced Best to make several significant retractions. Able to show the real timing of events, Macfarlane extracted from Best an admission that the insulin project was *not* effectively finished before Macleod returned from his holiday. Best had to concede that Macleod had indeed taken an active part in designing experiments and interpreting their results. Best was on much safer ground, though, in continuing to claim that he had independently prepared the insulin used in the first clinical trial, on Leonard Thompson. Moreover, in recommending that the documentary film end with the Thompson case, Best once more calmly skated over the fact that this first clinical trial was at the time considered highly unsatisfactory and regrettably premature. Its value to him lay in the chance it gave to write out of the story again the much more successful extract obtained and refined by Collip.

It did not take the Film Board long to realize that making their film would be bound to offend some very big names in Canadian science. The head of the National Research Council had no doubt at all who was to blame for this: 'it was a pity', he wrote, 'that Dr Best, a man of undeniably great gifts, had devoted so much time to building up his own part in the insulin discovery far beyond its actual importance'. They quickly came to see him as a major obstacle to making a realistic and historically worth-while documentary. In the event, despite a major preliminary invest-ment, the planned film was ditched and a highly emasculated version was shot instead. The film-makers simply refused to invest thousands more dollars propping up Best's fragile ego. Best's own reaction to this news is not known.

Although such events should have put Charles Best on his guard, he was by then in too deep. Almost certainly he had come to believe the half-truths and lies he was telling. This may have insulated him from what could otherwise have been a devastating blow. In 1954, powerful evidence against his own account surfaced in the *Journal of the History of Medicine and Allied Sciences*. Joseph Pratt, an American doctor, had sub-mitted a paper in which he explicitly argued that Collip's contribution had been imperative for the production of the 'first insulin ever to be used successfully in the treatment of diabetes'. Pratt concluded that all four team members had made useful contributions, but this did not change the fact that he had singled out Collip for special emphasis. William Feasby, Best's acolyte and biographer, wrote an irate reply, but this just raised more questions. A proposal was then made by Banting's biographer to publish in unison the early accounts of the discovery written by the main participants in 1922. This eminently sensible plan was soon scuttled by Best. He was now, quite understandably, feeling under siege. In a desperate move, he leaned on the President of the University of Toronto who forbade the publication of the relevant manuscripts that were by now the property of the university. Starved of oxygen, the fire temporarily died down.

Over the following years, more and more embarrassing evidence continued to materialize. Best had repeatedly claimed that Macleod had offered not a single morsel of useful advice. In 1957, his own papers gave the lie to this. To assist Feasby in writing his biography, Best found the summary of events he had written in 1922. This proved that several key purification procedures had only been performed 'with the benefit of Dr MacLeod's advice'. The crucial use of alcohol as a filtrate, it emerged, had been solely Macleod's idea. One especially important letter showed that a vital procedure Best had always claimed as his own—chilling the extract before purification—had really come from Macleod. Feasby's response was to ignore the material. Best tacitly approved of this, though his preferred reaction was much less cautious. Feasby had to talk him out of making full use of the letter, save only for the part prejudicial to Best's story.

Through all this, Best continued to use the public platform to cast aspersions against Macleod and Collip. They were accused of having stolen credit for his ideas and then taken advantage of Best's lack of powerful supporters to deprive him of proper recognition. 'If only I were not such a retiring Canadian', he once bemoaned.

Backdating the great discovery

Although Best always insisted that the Leonard Thompson trial had been a major triumph, one of his fundamental claims during the 1940s and 1950s was that the real discovery of insulin had been made during the earlier period when Banting and he were working alone with the depancreatized dogs. This must have seemed to him a brilliant strategy in that it left his audience to ask the $64,000 question. How could either Macleod or Collip possibly have made significant contributions if all the important work had been done when the former was still on holiday and the latter had yet to join the group? This was a line he tried to sell to the film-makers. In writing to Macfarlane, Best described how on Macleod's return to Toronto, 'There was no doubt in Macleod's mind when he looked over our data that we had the internal secretion of the pancreas'. 'You don't even have to mention Collip', he elaborated to a Commissioner of the National Film Board. In the same year, Feasby explained on behalf of his idol, 'convincing proof of the presence of insulin was available in the summer of 1921, when they [Banting and Best] were working alone and only on depancreatized dogs'.

This version of events was enshrined in Best's 1957 Oslerian Oration in London. There he enthused about 'The seven months of harmonious, intensive work which Fred Banting and I carried on together in 1921', which was 'the period which we both considered to be that of the Discovery of Insulin'. He then concluded his talk with the sentence: 'The picture of those days which I have tried to paint for you, is the one that I hope you will carry away in your minds.' These words have a curious— and revealing—tone of pleading.

Exactly as Best intended, many did go away believing that his failure to get a Nobel Prize was a travesty. With neither Banting nor Macleod around to object, several of those in attendance expressed their dismay that he had been deprived of his share in the prize. And as the encyclopaedia entry at the top of this chapter shows, the idea that the role of insulin was discovered in 1921 and that all Macleod and Collip did was refine a pre-existing proof gained wide currency. With many of the key documents unavailable, the one threat to Best's fantasies was Collip, and he chose to remain silent.

Given that there was no way in which Best could have been sure that he would remain so, there is something almost pathological about the way in which he remorselessly risked angering him into speaking. Remarkably, however, Collip kept his counsel until his demise in 1965. Even then there was no sealed envelope to be opened on his death, the contents of which would denounce his Nemesis. Collip's strongest reaction to the denigrating nonsense to which he had been exposed over the years was to tell friends that he was convinced the original papers would later speak for themselves. Although thanks to Michael Bliss, they have now done so, during his lifetime Best used all the considerable influence and power at his disposal to ensure that they, too, remained silent. To all intents and purposes, after Banting's death the history of the discovery of insulin had to rely on one primary source: Charles Best. Rarely has one man been in a better position to write his own eulogy. Even fewer have made such self-glorifying use of it.

A twist in the tale

There is, however, a twist in the tale that would have graced a story by Guy de Maupassant. By the time of Collip's death, Best's incessant self-aggrandisement was beginning to alienate even his closest friends. Feasby's book proved to be out and out hagiography, something that appalled Sir Henry Dale who joined with several others in suggesting that it be pulped. Given how much he had managed to distort the record, discretion should have been Best's preferred suit in the decade leading up to his death in 1978. Instead he approached publishers of texts dealing with the history of insulin and asked them to change their accounts of the discovery in his favour. He also embarrassed his friends by writing article after article implicitly accusing once-eminent physiologists of underhand behaviour, self-delusion, and crass selfishness. It was a pattern that seemed by now to have become central to his existence. But all the time, one of his earliest and most consistent distortions lay there waiting to ensnare him.

In the early 1970s, Ion Pavel, a Romanian physiologist, began to research a longstanding claim that Pavel's fellow-Romanian, Nicolai Constantin Paulesco, had beaten the Toronto team to it. The evidence Pavel uncovered proved conclusively that during the 1910s and 1920s Paulesco had been working on diabetes and the isolation of insulin. What's more, he had published the results of his experiments involving depancreatized dogs *before* Banting and Best had submitted their first paper in 1921. If we make the reasonable assumption that the Toronto team's real success lay in successfully administering insulin to human patients, the reputational threat posed by Paulesco's work was not very great. Yet Best had spent the preceding 25 years basing his claims of priority in discovering insulin on his and Banting's experiments with dogs. The implications were clear. If controlling the blood-sugar levels of dogs was adjudged the basis of the discovery, Paulesco's priority became extremely hard to deny.

This is the awful irony with which the elderly Charles Best was confronted. Ultimately his own manoeuvrings had been his undoing: not because his claims about 1921 and 1922 had been effectively and publicly disputed, but the very opposite. Best's reappraisal of the dog experiments had been so widely endorsed that many diabetologists sincerely believed this to have been the period of the actual discovery. As we have seen, this was a major distortion of events, the aim of which was to marginalize the first really effective clinical trials and Collip's crucial contribution to them. This underhand strategy now posed two problems. Not only did Banting and Best's paper come out after Paulesco's, but their work with dogs had been poorly carried out and produced what can best be described as inconclusive results. So in pushing his and Banting's mediocre experiments to the fore, Best was celebrating the least-satisfactory aspect of the work carried out at the Toronto lab. This meant that when Pavel began to investigate Paulesco's claims he was able to compare the Romanian's very creditable data with the worst any of the Toronto team ever published. Thanks to Best's machinations, there was a very real danger that, in an area of medical science central to Canadian national pride, priority would have to be yielded to Romania.

When full access was finally granted to the early Toronto papers, it became clearer than ever that Paulesco had performed essentially the same experiments as Banting and Best. The only differences were that he had started earlier and, by working with much more care, achieved greater success. It was also undeniable that the Romanian had published his data months before the Toronto team had even written their first paper. As one historian wrote in 1971, the Toronto team's work may be 'construed as confirmation of Paulesco's findings'. By the early 1980s, Michael Bliss argues, 'Paulesco's priority . . . was on its way to becoming a new orthodoxy in medical history and endocrinological circles'.

As he watched this starting to happen in the mid-1970s, Best had only one course open to him. Responding to a loaded question about Paulesco's work, he retorted that none of the Toronto team's rivals had 'managed to convince the world of what they had. This is the most important thing in any discovery. You've got to convince the scientific world. And we did'. Not quite the testament to his scientific career that he had spent the previous 50 years building up. And, as he faced up to the imminent prospect of death, Best must have been painfully aware that in so successfully writing Macleod and Collip out of history, he had come perilously close to destroying the whole Toronto team's claim to fame. The high profile that Macleod and Collip now posthumously enjoy on relevant Canadian websites strongly suggests that this is a lesson others, too, have learned.

Only human nature?

In commenting on what he falsely claimed were attempts by Macleod and Collip to usurp his glory, Best once remarked with an air of affected generosity, 'it is perhaps only human nature to claim some share and credit for procedures that have given important results'. It has frequently been observed that criticisms directed at others are often profoundly self-revelatory. If this is true of the remark made by Best, we have to assume that at some level he was aware of what he was doing. As this implies responsibility, we are led to ask how differently others would have behaved in similar circumstances. By the toss of a coin, he had found himself part of what first appeared to be a struggling duo, unlikely to achieve much success. Then, not least thanks to the pivotal contributions made by Macleod and Collip, it became a team that earned fame on a scale of which most scientists can only dream. The fairest assessment of what had happened to Best is that he had been amazingly lucky. He had some subsequent success in his research career, but not outstandingly so. Banting could have given him a larger measure of credit, but, after all, Banting was the senior man who had come up with the original idea. Further, it is now far from clear whether Best actually did anything deserving special credit.

It has also to be said that the individual members of the Toronto team provide such a spectrum of responses to extreme fame that Best's behaviour cannot be seen as typical given the circumstances. Aspects of Banting's behaviour mirror Best's, but with rather more cause. History might well have unfairly treated him as no more than an assistant to Macleod. Banting's diminution of Collip's role was distinctly Best-like, but there is now evidence that, immediately prior to his death, it was his intention to make amends. Macleod can be characterized as a man who took his full measure of fame, but then had the decency to leave it at that. Not for him the multiple rewriting of the historical record; nor did he yield to the temptation of engaging in a slanging match with his erstwhile subordinates.

The true antithesis to Charles Best, however, is James Collip. Now considered by many to have been the pivotal member of the team, he seems to have been wronged at every stage. Banting and Best's folly in testing their poor-quality extract on Leonard Thompson seems to have been explicitly intended to pre-empt any success Collip might have. Thereafter, although Macleod made the importance of Collip's role clear in his Nobel acceptance speech, Banting and Best worked assiduously to minimize his contribution. In the face of these repeated slights, Collip behaved with outstanding grace, confident in the knowledge that historical research would eventually vindicate him. It is therefore Collip who most clearly points up the extreme nature of Best's behaviour.

Few of us might be expected to face down gross provocation with the stoicism of Collip, but even fewer would have carried on striving for more and more glory with the same passion and unremitting dedication as Best did right up until his death. Nor would most of us be so self-deluding as to cling on to a raft of fabrications in blithe disregard of wave after wave of contradictory evidence. It seems likely that Michael Bliss is right in arguing that Best's distortions were the product of a somewhat unbalanced mind. Certainly he was hugely incautious in the stories he told. His manipulations were so serious and so shot through with contradictions that he found himself repeatedly driven to add embellishments, or change his story to meet each new set of discomfiting facts. In his final decade, rather than enjoying peace and serenity, he found he had painted himself into a corner. Generously given a quarter share in a reputational gold mine, he had tried to grab the entire claim and, in doing so, stood in imminent danger of losing it entirely.

Although most certainly not a typical example of the way in which scientists react to fame, Best is exemplary in the unintended warning he gives to others. Prolonged campaigns of self-glorification are not uncommon in science, particularly with regard to matters of priority. Collip represents the other side of this coin, showing how easily scientists who have made valuable contributions can be sidelined by those whose determination to achieve immortality is ethically unconstrained. Longevity also plays its part. With Collip silent and Banting and Macleod both dead within 20 years of their greatest triumph, the field was clear for Best to claim wisdom after the event and have such claims accepted by most of the scientific community. Others, too, have yielded to the temptation of reinterpreting experiments performed before major breakthroughs in the light of later knowledge. We saw a similar strategy in Joseph Lister's claims that he had always been a germ theorist in the style of Robert Koch and the Berlin team.

It does not seem to have worried either Best or Lister that ample evidence existed showing their supposed prescience to be no more than a figment of their imaginations. Ostensibly these were acts of extreme recklessness. To an academic, reputation is all; so why did these men hazard so much? Aside from their personalities, part of the answer must lie in the way in which their first forays into dishonesty were received. Had the intellectual environments in which either man was working delivered immediate and consistently negative responses, it is likely that they would have been far more constrained in attempting to tamper with the record. As this didn't happen, we can reasonably conclude that their stories were finding a ready market. With Lister, imperial Britain's wish for scientific heroes to match Pasteur and Koch must have been a major factor. It was an affront to what was then the world's wealthiest and most powerful nation to see all the cleverest medical science being done elsewhere. Here, then, were the ingredients for the development of a scientific myth surrounding Lister equating to Livy's tale of Horatio.

And much the same can be said of Best. To many, Canada may seem a near ideal country but Canadians have had a long struggle with their identity. Internally divided between the rival French and British connections, they also suffer from what some have likened to being in bed with an elephant: immediate proximity to the United States. Anybody who has read of the construction of the Canadian-Pacific Railway knows of the immense expense and trouble to which Canadians went to link their country with a railway running across Canadian soil (or ice-covered rock) throughout its length. Making use of pre-existing US lines for parts of the route would have been cheaper by millions of dollars, and much quicker. But relying on your neighbour for a crucial cross-continental link, did not seem the hallmark of a nation, so Canada went it alone. This was the environment in which the discovery of insulin took place. The United States was already starting to accumulate Nobel Prize winners and, more generally, its academic institutions were starting to rival the very best in Europe. Canada, with a much smaller population and still seen by some as a mere outpost of Empire, seemingly could not hope to rival it. Then, out of the blue, one of its handful of universities produced a medical break-through of world-class importance.

What the great and the good in Canada wanted was what Best wanted: to milk it for all it was worth. As a fond parent and taking vicari-ous pride in its stout sons of Empire, Great Britain wasn't going to cavil much about the contribution of the Scot, Macleod. Collip proved some-thing of a shrinking violet, so, at first, the Canadian Establishment threw all its weight behind Banting and his version of events. Then, on his death, they went for the even more high-profile Best. Whatever he said was fine as long as in aggrandizing himself, he kept Canada's finest scientific triumph continually to the fore. Given what their national film-makers discovered, there can be no suggestion that everyone was ignorant about what Best was about. But this was not of great concern. Where national honour is concerned, it would seem that the debate Graves imagined between Pollio and Livy is as relevant as ever.

Perhaps, though, those who follow Livy in subordinating historical truth to the national interest, should learn from the story of Charles Best what a dangerous game they are playing. For in the end, the story of Best's campaign for recognition is a personal tragedy and a national embarrass-ment. First, because it concerns a man so desperate for acclamation that he could never rest content with what he had achieved. Second, because the ultimate consequence of his machinations was to place in jeopardy not only his own place in scientific history, but also that of the entire team of which he was privileged to be a member. In the end, his country's primacy in this major field has only been preserved by that most salutary of experiences, rediscovering the truth.

ALEXANDER FLEMING'S DIRTY DISHES

> There can be no doubt of his [Fleming's] certainty that he
> was dealing with a matter of the greatest possible moment to
> medicine, even though nobody hastened to pile laurels on the
> cradle of penicillin or garland its discoverer, and though there
> was a danger that so valuable a truth might totter into ever-
> lasting obscurity.
>
> He tenaciously retained a firm, inner conviction that penicillin,
> 'would one day come into its own as a therapeutic agent'.
>
> L. J. Ludovici, *Fleming, Discoverer of Penicillin* (1952).

Ask a collection of British scientists to identify the most significant
chance event in the history of medicine and a sizeable majority is
likely to name Alexander Fleming's discovery of penicillin. They
will have in mind the broad outlines of an event that would later revolutionize medicine and save countless lives.

In September 1928, Fleming, a mild-mannered and seemingly reticent Scottish bacteriologist, returns from holiday to his laboratory in St
Mary's Hospital, London, and decides that it is about time that he cleans
the Petri dishes littering his workbenches. As he gathers them together, so
that they can be immersed in disinfectant, he is interrupted by a colleague
who asks him what he has been working on. By way of answer, Fleming
turns to his pile of dirty dishes and picks one at random that has escaped
disinfection. On examining the dish he mutters, 'That's funny'. It's one of
those British understatements (although uttered in a very different context, 'I may be some time' is another) that seem predestined for legend.
For in the dish that Fleming has selected, the bacteria experimentally
sown a week earlier have failed to colonize a region of the dish that is
home to a mysterious mould. The mould is immediately assumed to have

Left: Alexander Fleming (1881–1955) examining a Petri dish.

floated through the open window of Fleming's lab. But wherever it came from, it has impressive antibacterial properties.

From the first, Fleming senses that something very important has happened. He rushes into his colleagues' rooms bearing his Petri dish to show them his exceptional discovery. And within days Fleming announces that he has at last found a 'magic bullet'—the Holy Grail of bacteriology—which can fight infectious disease without harming the host. There follows years of further research and development, much of it performed in Oxford under Howard Florey, with Fleming's guidance and encouragement. Then penicillin is launched and rapidly emerges as the single most important therapeutic breakthrough in the history of medicine. By 1942, Fleming's initial insight has been spectacularly vindicated. The age of antibiotics is born and its inaugurator finally wins the credit his genius deserves.

Individuals born well after the events described may have difficulty fully grasping their real importance. But just consider for a moment that almost every family up until the late 1940s suffered some tragedy in consequence of bacterial infection. Many endured repeated losses. To generations among whom even mild respiratory infections and minor injuries posed a major threat to life, the coming of penicillin was therefore a gift from an unusually beneficent God. Worldwide, gratitude to the man who seemed to have so dramatically reduced the threat of bacterial infection was unsurprisingly immense. Cannoned into international celebrity, Fleming subsequently received 25 honorary degrees, 26 medals, 18 prizes, 13 decorations, the freedoms of 15 cities and boroughs, and the membership of 89 academies and societies. He was also granted five private meetings with the Pope.

Although it might be suggested that by so enthusiastically endorsing this story, the scientific community is elevating good luck over good science, this would be to misread the morals that can be drawn from it. Scientists happily accept that Fleming's discovery of the antibacterial properties of penicillin was highly fortuitous. But they consider him an outstanding scientist because of his immediate recognition of the importance of a few blotches in a Petri dish. Fleming seemed instantaneously to know that he was on to something. In contrast, all his colleagues could muster was polite interest. This one act, then, perfectly exemplifies the foresight and observational skills scientists consider central to their profession. When showing his colleague a randomly selected Petri dish, Fleming had the genius to see what others would have ignored. He was not distracted by thoughts of his recent holiday. His interest had not been dulled by the years he had spent 'toying' with bacterial colonies without major success. As one scientist has said (quoted in J. L. Ludovici's biography of Fleming), 'Though Lady Luck may fly in at the window, she has a way of eluding the undiscerning—believe me!—and settling where she's not unlooked for'. When opportunity came knocking, Fleming was ready. And that is why he is an icon of science.

But as with the other case studies in this book, the difficulties start when we take a closer look at what actually happened. The facts surrounding Fleming's discovery and the subsequent development of penicillin prove to be far more complex than the myth that I have just outlined.

The standard account makes four major claims. First, that the events leading up to Fleming's discovery of penicillin involved a lion's share of luck. Second, that this was the first time the antibacterial properties of the mould *Penicillium* had been noticed. Third, that Fleming was quick to grasp its therapeutic importance in curing infectious disease. Finally, that he worked hard to realize the potential of penicillin as a mass-produced wonder drug.

Close examination of these claims, however, enables two general points to be made: first, Fleming's recollections of the crucial years 1928–42 indicate a memory of the most partial and unreliable kind; and, second, Fleming was perhaps the luckiest *and* the unluckiest man in the history of science. Some readers may well be familiar with these claims, but for no other reason than that the evidence has taken a long time to emerge, the conventional portrait of Fleming's role in the discovery of penicillin retains a very wide popular currency. Relying largely on the research of the historian Gwyn Macfarlane, I try in this chapter to prize apart the man and the myth.

Fortune and fakery

In 1906, Alexander Fleming joined the Inoculation Department of St Mary's Hospital in London. The department was run by Almroth Edward

Wright, a man with a self-consciously brilliant mind and an infamously acerbic wit. Under Wright's direction, Fleming's principal role was to identify pathogenic organisms, grow them in culture, kill them, suspend them in a suitable fluid, and administer them as vaccines. His first significant success was in 1909 with the preparation of a vaccination against acne. In the same year, the German Paul Ehrlich discovered that an arsenic compound, dubbed Salvarsan, could fight the syphilis spirochaete within the body without adversely affecting the host's tissue cells. This was an outstanding example of the kind of 'magic bullet' for which numerous bacteriologists hunted in vain. Ehrlich happened to be a close friend of Almroth Wright and he gave St Mary's a monopoly for producing and injecting Salvarsan in Britain. Wright generously passed the opportunity on to Fleming, whose private practice and public reputation soon owed much to the promiscuity of Edwardian England. It was this experience, he would later suggest, that made him acutely aware of the possibility of finding more chemicals that could eliminate bacteria within the human body. Thus he was prepared for the two major chance events that defined his career.

The first occurred in January 1919. Fleming was suffering from a cold. His sinuses were producing large amounts of mucus that would have seemed no more than bothersome to anyone else. Fleming, in contrast, saw it as a ready source of bacteria for his experiments. Repeatedly scraping the inside of his nostrils, he spread the contents on his ever-ready Petri dishes, lined as always with a nutrient medium, and left the bacteria to develop into colonies. The results of one dish in particular were perplexing. A small colony of bacteria had grown on the dish but in the vicinity of the nasal mucus itself the dish was entirely clear. To Fleming, the implication was obvious: something in the mucus had killed the offending bacteria. He rapidly deduced that bodily fluids contain a natural antibacterial agent and tests using all manner of human and animal fluids confirmed this presupposition. He had inadvertently discovered the enzyme now know as lysozyme.

But, as later proved to be the case with penicillin, Fleming had been exceptionally lucky with his antibacterial agent/bacteria combination. The bacteria that had colonized the Petri dish was of a kind almost uniquely susceptible to being eliminated by lysozyme. Named 'A. F.

Coccus', this strange and rare microbe facilitated a discovery that might otherwise have been delayed for decades. At first, Fleming thought that he might also have discovered a new way of fighting disease. To his grave disappointment, however, he was soon forced to admit that the toxicity of lysozyme is confined to non-pathogenic microbes. Consequently it is useless in the clinical setting. Nevertheless, writing many years later, Fleming explained that his discovery of lysozyme did do him the immense favour of alerting him to the possibility of identifying more naturally occurring antibacterial agents.

The second great chance event that shaped Fleming's professional life was his momentous discovery in September 1928. Indeed, we now know that Fleming was even luckier than is usually supposed. The *Penicillium notatum* that appeared in Fleming's Petri dish is exceptionally rare and has far more potent antibacterial properties than any other kind of *Penicillium* mould. The chances against this particular species settling on his Petri dish were astronomically high. In the hours and days after the mould landed, however, even higher odds were quickly racked up.

As described above, Fleming had left a Petri dish out of the way in his laboratory while he went on holiday. On his return, an ex-colleague (Dr D. M. Pryce) was shown a Petri dish containing a mould that seemed to have killed several colonies of bacteria. In his published account, Fleming explained that he had left this particular dish covered with bacteria before he put it to one side. From this the order of events seems clear. First the bacterial colony was grown and then mould spores landed on the dish and eradicated any germs within their immediate vicinity. A colleague then identified the mould as *Penicillium* and Fleming repeated his experiment with several pathogenic bacteria with success in nearly every case.

Yet, Fleming's attempts to reproduce his experiment consistently failed. In the days after he had murmured 'That's funny', his experiments showed that *Penicillium* mould can grow among bacterial colonies but leave them entirely unaffected. His *Penicillium* seemed to have lost its killer instinct. Eventually, Fleming found that to re-obtain the result that had first attracted his attention he had to reverse the procedure. First, the mould had to be grown at the room temperature it finds ideal. Then, staphylococci bacteria had to be sown near it and the dish placed in an incubator at a temperature suiting bacterial growth. When this procedure was followed, the bacterial colonies within 3 centimetres of the penicillin mould died exactly as they had in Fleming's original dish. The mechanisms underlying this finding were not fully revealed until 1957. James Park and Jack Strominger then found that *Penicillium* does not, as Fleming thought, penetrate bacterial cells and explode their contents. Instead, it interferes with bacterial cell division, preventing the synthesis of compounds needed to build bacterial cell walls. So once a bacterial colony has formed, *Penicillium* is ineffectual.

Given this constraint, the true magnitude of Fleming's luck can soon be perceived. To get his first chance result, he must have sown his dish with staphylococci before going on holiday and left it unincubated on his bench. Fortunately for him, as records show, the weather for the following few days was cool and did not favour the growth of the bacterial colonies. In the meantime, a rare *Penicillium* spore flew up, probably from the mycology section on the floor below. The spore landed on Fleming's Petri dish during this cool spell. At a low temperature, suiting its growth, the mould established a healthy colony and began producing its antibiotic product: penicillin. Then, again as records show, the weather changed. A warm period stimulated the growth of the staphylococci, which then formed colonies everywhere on the dish except near the mould. There the newly forming bacteria simply couldn't form proper cell walls and were destroyed in the attempt. Clearly, then, Fleming's discovery of penicillin was built on an impressive edifice of chance events, several of which had an exceptionally low likelihood of occurring even in isolation.

It tells us something about Fleming the man that he never admitted to the fact that he needed to sow the *Penicillium* mould first if its antibacterial properties were to be demonstrated. When he published his findings in 1929 researchers seeking to replicate his findings had to work out for themselves that, under normal circumstances, the order of events suggested by him needs to be reversed. Perhaps in order to preserve the simplicity of the original account, after 1928 Fleming adopted a strategy of modest deception. Much of his data, as presented in his 1929 article, is in this limited sense faked. And although he had every reason to believe that penicillin *can* destroy pre-existing bacterial colonies, his failure to admit to having experienced difficulties in replicating his first 'experiment' rendered it much harder for other bacteriologists to prove the efficacy of penicillin for themselves. But this is perhaps the least significant flaw in the conventional story.

'A very well-known phenomenon'

The other central character in this drama is the Australian-born pathologist Howard Florey. In 1929, he followed in Fleming's footsteps by investigating the nature and therapeutic potential of lysozyme; within 11 years he would make the crucial breakthrough in the development of penicillin and, as a result, share the Nobel Prize. Publishing an account of his work with lysozyme in 1930, Florey commented in the appendix that one bacteria inhibiting the growth of another was 'a very well-known phenomenon'. In support of this claim he cited a book published in France in 1928 called *Les associations microbiennes*—before Fleming's lucky observation—by George Papacostas and Jean Gaté. This book underscores the fact that if Fleming's fame was based purely on his discovery of the antibacterial properties of the *Penicillium* mould, his status would be very hard to defend. This is because Papacostas and Gaté were able to cite more than half a dozen eminent scientists, in Britain and abroad, who had previously noted and investigated the clinical possibilities of *Penicillium* variants.

In 1875, for example, the famous British physicist John Tyndall described to the Royal Society in London how a species of *Penicillium* had made several bacteria burst open and die. Joseph Lister had even made clinical use of it: in 1872, he noted, 'Should a suitable case present, I shall endeavour to employ *Penicillium glaucum* and observe if the growth of the organisms be inhibited in the human tissues'. Such an opportunity did arise, and he reported using the *Penicillium* to great effect on a patient injured in a road accident. It remains unclear why he did not follow up this apparent success. Some years later, on the Continent, *Penicillium* research was taken an important stage further. In 1897 a young French army doctor, Ernest Duchesne, described in his doctoral dissertation the effectiveness of *Penicillium glaucum* in animals injected with normally fatal doses of pathogenic bacteria. In a subsequent article, Duchesne emphasized the potential therapeutic value of *Penicillium*. Tragically, however, he died of tuberculosis before he could make further progress. *Penicillium* research was not extended past the point reached by Duchesne for another 40 years. As will be seen, even though Duchesne was long forgotten, Fleming's personal contribution never advanced beyond it.

For the numerous bacteriologists and mycologists familiar with the work of Duchesne, or the book by Papacostas and Gaté, there was nothing especially striking about Alexander Fleming's article of 1929 in which he announced his lucky 'discovery'. Two circumstances, however, helped to ensure that the mantle of 'discoverer' was bestowed on Fleming and not on Lister, Tyndall, or Duchesne. First, when the full-fledged Fleming myth first took off in the 1940s very few people cared to remember what long-deceased rivals may or may not have achieved. Second, it had become obvious that Fleming's *Penicillium notatum* was very much more potent than the mould strains used by any of his predecessors. Had Lister, Tyndall, or Duchesne been lucky enough to capture a spore of this mould, penicillin might have been developed decades earlier. The luck, however, was all Fleming's. But far more important than either of these reasons is another. Fleming always claimed that he had realized from the first that penicillin would emerge as a wonder drug. And during the 1940s those contributing to the making of the myth never openly questioned this claim. With the benefit of careful historical research, however, we can now turn to the crucial question of whether Fleming's version of events is actually supported by the facts.

'My old penicillin'

The popular image of Fleming's role in the advent of antibiotic medicine has an uncanny tendency to collapse time. The gap of almost 15 years between the immortalized Petri-dish incident and the development of a clinically usable drug is either glossed over or put down to the notoriously long years of R & D that drug companies habitually refer to in justifying the cost of their products. If the delay is thought about at all, the image floats into our minds of years consumed by unavoidable work with animal models, clinical trials, and product improvements. But, from the outset, the length of this period in the case of penicillin was strenuously downplayed. In Britain, the wartime press quickly seized on the idea of penicillin as a new wonder drug. And, in articles with headings such as

'A Vital Discovery' and 'The Cure that Came Through the Window', journalists presented a seamless transition from an autumn day in 1928 to the first successful clinical trials of penicillin in 1941. From these news-paper accounts few readers would have guessed just how long separated the two events described.

Furthermore, most of the articles celebrating Fleming's efforts did not even mention the names of the men at Oxford University's William Dunn School of Pathology—Howard Florey, Ernst Chain, and Norman Heatley—whose technical breakthroughs enabled the mass production and proper clinical testing of the new drug. As Fleming was honoured with an ever-rising number of awards and Papal meetings, the Oxford group became increasingly disgruntled. Unfortunately for them, the initial press exposure set the trend and Fleming was permanently lodged within the popular consciousness as the discoverer of penicillin.

Yet close examination of the record now suggests that had the accolades for the discovery and development of penicillin been allocated on merit, Fleming would not have made the short list. At first glance, this might seem to be a ludicrous claim. But there are several sound reasons for accepting it. The first arises from re-appraising the scale of the contri-bution made by Florey's Oxford team. Between 1938 and 1941, they succeeded in producing enough penicillin to perform a series of startlingly successful clinical trials on human patients. Then, working around the clock, they made the numerous technical advances that were essential if penicillin was to be produced on the scale demanded by the war effort. Without any assistance from Alexander Fleming—one of them even believed that he was dead—they advanced penicillin research from little more than a raw discovery to the threshold of a mass-produced wonder-drug in a mere 3 years. So much for the 15-year struggle implicit in the legend.

Fleming got into contact with Florey's team only in 1941, immedi-ately after they published some impressive clinical results in *The Lancet*. Then, showing like Mark Twain that reports of his death had been grossly exaggerated, he went to Paddington Station and caught a train to Oxford. Arriving at the Oxford laboratory with little notice on a Monday morning, he shook hands with Florey and remarked with affected nonchalance, 'I've come to see what you've been doing with my old penicillin'. It was an apparently offhand remark that Fleming had probably been rehearsing ever since he had read of their success, and it immediately raised hackles. To Fleming it served to stake a claim. But if he thought Florey's team would immediately bow and scrape, he had badly misjudged the situation. After years of labour in difficult circumstances, Florey's team felt strongly proprietorial towards this strange 'mould-juice'. Still, Fleming's visit passed off without further unpleasantness and no one foresaw the bitterness that would later ensue.

Soon after Fleming's excursion, the *British Medical Journal* carried an editorial on penicillin, which, after noting Fleming's initial discovery, went on to assert that the true clinical potential of the mould had been recognized only by Florey's team. Fleming was horrified. He immediately fired off a rebuttal in which he cited extracts from his published papers to prove that he had always believed in the therapeutic value of penicillin. Referring to an article of his in the *British Dental Journal* of 1931, he quoted, 'Penicillin is valuable to us at present in the isolation of certain microbes, but it is quite likely that it, or a chemical of similar nature, will be used in the treatment of septic wounds'. Concluding his letter, Fleming wrote that the Oxford team had made important advances that 'enabled a clinical trial to be made which was more than justified by the suggestions I made ten or more years ago'. This is the version of events now generally accepted. The question is, should it have been?

'The period of failure and neglect'

Giving due credit to Florey's team is not the only reason for 'downsizing' the scale of Fleming's contribution. A second factor arises out of a major difficulty that has confronted his many biographers. Most have accepted their hero's claim that Florey's research was a natural progression from the work Fleming had carried out at St Mary's in 1928 and 1929. In doing so, they have all been forced to wrestle with two very uncomfortable facts: one general and one specific.

The general problem is that far from enthusiastically following up his chance discovery of 1928, Fleming largely stopped his research into the clinical value of penicillin within months of his original insight. The specific problem is that Alexander Fleming never tested the efficacy of his penicillin supplies on animal models infected with harmful bacteria. If a medical scientist thinks that a compound may be therapeutically useful, the obvious next step is to inject it into animals and then infect the hapless creatures with pathogenic bacteria. If enough animals recover from the second injection, then one may reasonably move on to human patients. Fleming's failure to perform this test—even though Florey's team did so when their state of knowledge was not far advanced from his in 1929—is of absolutely vital significance. For although Fleming did chance on the crucial mould, this is a quite different achievement from having fully realized its therapeutic implications. Indeed, his failure to undertake work with experimental animals implies very much the reverse.

How do his biographers skate round this? In their accounts of these awkward 12 years between 1929 and 1941, most have met the challenge head-on. Chapter headings such as 'The evil of waiting' (Lorenz Ludovici) and 'The period of failure and neglect' (Stanley Hughes) typify their attempts to suggest that Fleming's ambitions to develop the new wonder drug were dogged by technical difficulties, ignorance, and obstructionism. A frequent claim is that Fleming was deterred from further developing penicillin as a consequence of his natural shyness and modesty. Even though he realized that he had found something very important, he was not given the encouragement he needed to pursue his humanitarian aspirations. This is, however, extremely improbable. Fleming published 27 articles and papers between 1930 and 1940 in which reference to the therapeutic possibilities of penicillin could have been made had he thought the issue worth discussing. In fact, it is scarcely touched on. In a 1931 lecture entitled 'The intravenous use of germicides', he did not so much as mention penicillin, instead predicting that mercury-based compounds represented the most promising means of treating infectious diseases.

After 1929, Fleming made only one fleeting public reference to penicillin, and then his focus was certainly not on the treatment of internal infection. As his letter to the *British Medical Journal* made clear, in 1931 he told the readers of a dental journal that penicillin 'is quite likely [to be] used in the treatment of septic wounds'. But far from being but one instance amongst many, this reference is an isolated example. It is not much of an overstatement to say that by quoting it in his letter, Fleming was seeking to pass-off what had been little more than a footnote, as a lifetime's work.

Despite what some biographers have claimed, personal diffidence cannot explain the extreme paucity of references to penicillin in Fleming's publications. For somebody so handicapped, publication offers the ideal means of getting a novel idea across. Yet Fleming did not exploit the opportunities professional journals afforded. This strongly suggests that he had already said all he had to say about penicillin—and that amounted to very little. In any case, even if Fleming was a somewhat taciturn, dour person, he was never actually shy. The man who seldom shunned the opportunity to present his work publicly, happily performed in amateur dramatic productions dressed as a woman, and allowed dozens of Spanish well-wishers to kiss his robes during a procession after receiving an honorary degree, is not likely to conceal what he feels to be a momentous scientific discovery. It just will not do to suggest that Fleming kept his belief in penicillin to himself because of crushing shyness.

Another strategy his defenders have adopted has been to argue that Fleming's lack of confidence was exacerbated by the short-sighted hostility of his peers. One biographer, André Maurois, has described how Fleming announced his penicillin findings in 1929 at the Medical Research Club in London only to meet a roomful of blank and apathetic stares, 'the icy reaction to something which he knew to be of capital importance appalled him'. For all the shades of Gregor Mendel and the Brno meeting room, the harsh facts are that Fleming was a notoriously bad speaker, apparently incapable of altering pitch and emphasis in making presentations, and was often unable to make himself heard across a large room. But presentational problems were not the real cause of his failure to get his message across. There was a major substantive limitation: for reasons explored below he did not even mention the therapeutic implications of his discovery in front of the 1929 Medical Research Club meeting.

An alternative approach to explaining Fleming's supposed 'wilderness years' has been to accuse Almroth Wright, his head of department at St Mary's, of standing in the way of further penicillin research. 'All Wright's instincts were up in arms against penicillin', one biographer has written. This claim has little foundation. Although Wright was generally sceptical about the use of antiseptics in curing internal infections, during the mid-1930s he gave plenty of support to Fleming in his experiments with the most-important breakthrough in chemotherapy before the advent of penicillin—the development in Germany of the dye-related compounds called the sulphonamides. It has also to be borne in mind that, by the 1930s, Wright was a very old man who was far more concerned with developing recondite philosophical theories than standing in the way of a younger colleague with a potentially important idea. It is true that when Fleming did discuss penicillin as a therapeutic agent with St Mary's clinicians, he elicited very little enthusiasm for clinical trials. This, however, was primarily due to the clinicians' frustration at an Inoculation Department that seemed to prepare one supposed therapy after another, hardly ever achieving an encouraging result. Even so, had Fleming pushed harder, it is likely that he would have enjoyed more success. The crucial point is that he didn't push. Why?

Significantly most biographers have also tried to argue that Fleming was consistently let down by the biochemists to whom he had entrusted the responsibility of identifying and purifying the active ingredient of *Penicillium* 'mould-juice'. Fleming, it is claimed, saw this as the only means of producing penicillin in large-enough quantities to permit proper clinical trials and, ultimately, large-scale medical usage. But to his considerable chagrin, several teams of biochemists are said to have prematurely abandoned their researches. With sloppy biochemists being deterred by the smallest of obstacles, the world had to wait till the early 1940s before enough penicillin could be purified and its usefulness properly demonstrated. Writing in 1946, Fleming joined in this frenzy of biochemist-bashing: 'I had failed to advance further for the want of adequate chemical help . . . the problem of the effective concentration of penicillin remained unsolved.' Yet, Fleming's insistence that biochemistry halted his progress, leaving him wringing his hands in frustration for the want of a competent midwife for his wonder drug, is an elaborate fiction that did his biochemist colleagues a great disservice.

From wonder drug to reagent . . .

Fleming's discovery of lysozyme was accompanied by feelings first of rapture and then of disappointment. Although this enzyme represents a death sentence to certain rare microbes and does not harm body cells when injected into living organisms, it does not adversely affect pathogenic bacteria. As Fleming no doubt reflected with a feeling of resignation, his early experiences with penicillin followed a similar path. After showing his Petri dish to Dr Pryce, he was elated. Realizing that here was something of genuine importance, he photographed the dish and rendered it permanent by exposing it to formalin vapour. Once he had mastered the necessary techniques, follow-up tests were also encouraging. These showed that the *Penicillium* 'mould-juice' was effective against most dangerous bacteria in Petri dishes. It was ineffective only against a few types of bacteria, notably 'Pfeiffer's bacillus'. A highly significant bacterium, now called *Haemophilus influenzae*, this was then mistakenly thought to have been the cause of the deadly influenza outbreaks that killed more than 20 million Europeans in 1918 and 1919 (it actually causes meningitis and some other serious infections). But given its otherwise exceptional spectrum of effectiveness, Fleming did think at this stage that in penicillin he had the 'perfect antiseptic'. So he set in train the experiments that begin any new compound's journey on the way to use in the clinical setting.

He first tested its effects on mixtures containing blood serum, to see whether penicillin could discriminate between bacteria and body cells. To his satisfaction, the blood cells survived the experiment unharmed. At this point, however, doubts began to creep in. The 'mould-juice' had taken several hours to kill bacteria in the serum. Moreover, within only a few hours the penicillin had lost almost all of its germicidal power. Putting these concerns to one side, he then injected penicillin into healthy experimental animals. The animals suffered no adverse side-effects. But assays of the animals' blood showed that the penicillin had lost all of its therapeutic power within minutes.

Fleming had previously demonstrated that it required 4 hours to kill bacteria outside the body, so he was led to concede that the therapeutic value of penicillin was very highly circumscribed. More than a decade later, doctors and nurses would encounter the same problem. Initially the effects of penicillin on infected wounds would be dramatic; but so much was needed to destroy infection completely that medical staff were frequently reduced to tears when the penicillin supply ran out and the patient died. Even re-using what could be filtered out of the patient's urine rarely made good the shortfall. In the end, high-volume production would overcome the problem. But in 1928, not unreasonably, Fleming felt that his penicillin research might all have been in vain.

Fleming retained his enthusiasm for just long enough to recruit two young biochemists to tackle the problem of producing pure specimens of penicillin. Frederick Ridley and Stuart Craddock began their work early in 1929. Nevertheless, signs that Fleming's interest was flagging were already in evidence. Accordingly, he made no effort to find his bio-chemists proper laboratory facilities. To the inconvenience of the rest of the staff using the building, they were installed on a bench in a corridor outside the toilets. Given that both were prop forwards in the St Mary's rugby team, their physical bulk may well have exacerbated the problem. On the other hand, it may have served to discourage criticism.

Despite their less-than-ideal conditions, Craddock and Ridley made good progress. Crucially, before long they had produced quantities of the 'mould-juice' of a strength similar to that which Florey's team would later use in their clinical trials. Craddock and Ridley also showed that by keep-ing penicillin in a mildly acidic solution, and at a low temperature, its activity could be preserved for several weeks. Realization of this gives the lie to the claim Fleming made many years later that penicillin could not have been effectively developed as a drug in 1929 because of its 'instability'. In later life he also promoted the prevalent belief that the recent invention of freeze-drying had allowed the Oxford team to make its breakthroughs after 1940. Both claims are untrue. Fleming's team had enough penicillin in 1929 and sufficient knowledge of how to preserve it for them to perform animal experiments. Had they done so they would almost certainly have gained the confidence to go on to develop penicillin as an intravenous cure. But Fleming, Craddock, and Ridley did not bother testing penicillin against a real infection in animals because they assumed it would be a waste of time.

During 1929, Craddock and Ridley had reproduced Fleming's tox-icity studies on experimental animals and they, too, found that penicillin lost its germicidal powers in a frustratingly short period of time. Fleming greeted their results without surprise. He had shown scant interest in their research over the previous months and his fleeting comment on their work in his 1929 article on penicillin hardly did them justice. In the meantime, his few attempts to use penicillin in the clinical setting delivered highly ambiguous results. Assuming, on sound clinical grounds, that 'mould-juice' would be useless in treating internal infections, Fleming experimented with it as a local antiseptic. His first patient was a man who was dying of septicaemia after the amputation of his leg. The wound was irrigated with penicillin but the infection was hardly affected and the man soon died: not an auspicious start.

Next, one of Fleming's colleagues on the St Mary's shooting team developed a serious case of pneumococcal conjunctivitis. With the possible consequences for his shooting club to concentrate his mind, Fleming applied penicillin to the infected eye. The infection cleared up almost immediately. Nevertheless, this highly gratifying result was soon neutralized by Craddock's finding that penicillin was of no value at all in treating the persistent sinus infections from which he suffered. Thus, despite one or two successes, by the summer of 1929, Fleming's interest in penicillin was at a low ebb. Where the 'mould-juice' worked, there were clear alternatives, and on unimpeachable scientific grounds he denied that it could ever be the 'perfect antiseptic' he had briefly anticipated. Fleming's second slow descent back to earth was completed.

During 1929 he presented just one lecture and one article on his penicillin work. His lecture was the one referred to above at the Medical Research Club. In stark contrast to the popular myth, he did not introduce to a pathetically apathetic audience the results of experiments in the use of penicillin as a wonder cure. Instead, his paper was entitled 'A medium for the isolation of Pfeiffer's bacillus' and concentrated on altogether more prosaic themes. Because Pfeiffer's bacillus was of such enormous interest to clinicians as a potential cause of influenza, it was of profound importance to St Mary's Inoculation Department as the possible basis for a much sought-after flu vaccine. Expose a Petri dish containing *Penicillium* mould to a germ-rich atmosphere and in a few hours one can have pure colonies of Pfeiffer's bacillus with which to produce vaccines. This is because by eliminating almost all other bacteria, penicillin leaves the field clear for Pfeiffer's bacillus, one of the very few unaffected by it. No doubt this was a very useful innovation. But it was hardly the sort of discovery to win Fleming a standing ovation from a room of highly accomplished medical scientists with very diverse interests. When coupled with his lacklustre lecturing style, a somewhat embarrassed silence hardly seems surprising.

Fleming's May 1929 paper was not much more prescient. Entitled 'On the antibacterial action of cultures of *Penicillium* with special reference to their use in the isolation of *B. influenzae*', it described his lucky discovery but made only a passing reference to the therapeutic use of penicillin, and then only as a possible 'local antiseptic'. Thereafter, until the early 1940s, Fleming only once discussed penicillin's potential medicinal properties in print. Its importance for him was as a laboratory reagent used in isolating Pfeiffer's bacillus: that was all. (Fleming also used his 'selective weed-killer' as a routine diagnostic test: he wiped the sputum of patients with chest or throat infections onto a dish containing penicillin mixed into a culture medium, and waited to see if colonies of Pfeiffer's bacillus would develop.)

Fleming undertook no further penicillin research in the summer of 1929 and quickly terminated Craddock and Ridley's project. His laboratory notebooks of 1930 contain not a single reference to penicillin. A few pages are devoted to it in 1931 and 1932, and during the following 4 years penicillin was mentioned on just four occasions, in every case in relation to its role as a laboratory reagent. Then, in December 1938, he noted how penicillin can be produced in a variety of organic media.

Fleming's interest in the therapeutic properties of penicillin did occasionally re-surface. Thus, some of his colleagues later recalled him prescribing penicillin to treat patients with boils. But, again, his use of the mould was confined to treating external conditions and he still refrained from proper animal testing. So even if Fleming didn't exactly forget about penicillin, as some historians have claimed, there can be little doubt that had Howard Florey not hung tenaciously on to the conviction that penicillin is therapeutically viable, Fleming himself would never have seen it developed as a cure for internal infection. He was lucky in September 1928, but he was even luckier in having Florey and his team as his successors.

The biochemists have a go

Fleming's lack of interest in penicillin is apparent from his relationship with the few biochemists who selected it as an interesting project during the 1930s. Irrespective of what later biographers claimed, these men investigated the mould without the bidding of Alexander Fleming. And despite numerous solicitations, they also worked largely without his support. The first independent team to take on penicillin comprised the British biochemists P. W. Clutterbuck, Reginald Lovell, and Harold Raistrick. In 1931, Lovell telephoned Fleming several times to ask for bacteriological advice but the latter never once told him of the important biochemical work of Ridley and Craddock. Eventually, noting that the mould-juice had strange biochemical properties that would require an enormous amount of time to investigate—probably for very little gain— this group abandoned the project.

Fleming later claimed that this sort of 'failure' had caused the impasse that prevented him from developing a penicillin drug himself. This was largely wishful-thinking. After several months of dedicated labour, Clutterbuck, Lovell, and Raistrick managed only to move slightly further than had Craddock and Ridley under Fleming's lax supervision. And had Fleming only told Clutterbuck and his colleagues of the unpublished work in his department performed during 1929, he could have saved the biochemists months of fruitless labour. Furthermore, although he was well aware of their efforts, he never once gave Clutterbuck, Lovell, and Raistrick the sort of encouragement consistent with his later claim that he had passed the baton on to them to find a way of producing large quantities of pure penicillin.

As we have seen, by 1930 Fleming had only the most oblique interest in penicillin. But in 1934 another biochemist, Lewis Holt, joined Almroth Wright's team at St Mary's to work on the cause of scurvy. In a part-time capacity, he was also attached to Fleming's research team. Holt was intrigued by the power of penicillin and with some mild encouragement from Fleming he sought to extend the progress of Raistrick and his colleagues. Again, however, Fleming failed to mention the work Craddock and Ridley had undertaken. After making some early progress, Holt grew discouraged by the instability of penicillin, by the lack of enthusiasm expressed by Fleming, and by Wright's belief that he was wasting his time. Eventually Holt was moved to another project. In these months, Fleming hardly presented himself as the anxious 'father of penicillin' he was portrayed as in later legend. Even when biochemical research on penicillin was being undertaken on a bench close to his own, his interest was firmly fixed elsewhere.

If we search for a motivational commonality in this, the dog in the manger comes to mind. Was it that having unsuccessfully struggled with penicillin himself, he felt no inclination to smooth the road for others? In any event, by 1935 he had become a determined advocate of the combined use of vaccination and sulphonamides in treating disease. And when Florey and his Oxford team began publishing intriguing results on the utility of penicillin, he persisted in using public lectures to recommend this dual approach. In 1941, there was every danger of him being left in the dust by a team that had developed his own initial discovery.

Indeed, just when Fleming's help was most needed to promote the large-scale production of penicillin, he was almost entirely inactive. By August 1941, Florey's team had shown the mould-juice to be of enormous clinical potential. But they were struggling to persuade the major pharmaceutical companies to embark upon industrial-scale production. The Wellcome organization, Boots, ICI, and the Lister Institute, already overstretched by the war effort, were now being asked to invest large amounts of money in the production of penicillin. But they were acutely aware of the possibility that having built expensive plants, they would be rendered obsolete by a biochemist finding a cheaper and easier way of artificially synthesizing the active compound. Did Fleming thereupon leap into the fray, crusading for his brainchild? Certainly not. It seems that he still did not perceive the full therapeutic potential of a compound that he had personally discovered. Not until he had seen its power for himself in curing a family friend did he attempt to re-enter a field he had abandoned almost as soon as he had opened it up.

Reflecting on why Fleming failed to test penicillin on experimentally infected animals, one of the most brilliant members of Florey's team, Ernst Chain, baldly declared that doing so had not occurred to him. This is an understandable remark coming from a man who felt that Fleming had unfairly monopolized the accolades for discovering penicillin. But it is unlikely to be correct. True, had the St Mary's team used the amount of penicillin available to them in early 1929 to protect a mouse from bacterial infection, it is probable that they would have gone on to develop the wonder drug. But their early demonstration of the speed with which penicillin loses its potency convinced them, quite legitimately, that they should look elsewhere for the 'perfect antiseptic'. Given the information then available to them, there was nothing irrational in their concluding that experiments with infected animals would have been a waste of time and effort.

With luck, Fleming, Craddock, or Ridley might have taken that crucial further step; but, for once, luck wasn't on Fleming's side. As a result, by the summer of 1929 he had no reason to think that penicillin would outperform any of the other antiseptics available to clinicians. Moreover, by 1935 the advent of sulphonamides quite plausibly suggested to him that dye-stuffs and not moulds represented the future of clinical bacteriology. Thus, Fleming's failure to encourage biochemical research or to win the support of St Mary's doctors was the result of a lack of interest in penicillin as a cure. It was certainly not the effect of excessive modesty—something that his subsequent history strongly suggests was not one of his personal defects.

Account also needs to be taken of one other factor. Although Fleming had been alert to the direct therapeutic possibilities of both lysosome and penicillin, his chief employment was as a leading member of an Inoculation Department engaged in the development of vaccines. Even when his own therapy-directed efforts had run into the sand, penicillin remained of considerable usefulness as a laboratory reagent in his vaccine work. As its discoverer, that alone would have been reward enough for Fleming— until, that is, the Florey team's success revealed that he had missed the opportunity of a lifetime.

'This great pioneer'

As we have seen, when Fleming wrote to the *British Medical Journal* implying that Florey and his colleagues had merely realized his aspiration of 1929 he was stretching the truth way beyond its limits. Yet, this is a claim that his colleagues and biographers accepted with alacrity. Between 1942 and 1945 Fleming became one of the most famous men in the world, seen as the outstanding example of the modest, unassuming, but observationally brilliant scientist. The Oxford team that had actually developed a pure form of penicillin and proved its therapeutic value felt understandably aggrieved. Although Fleming never once claimed credit for the work performed in Oxford, he never needed to. The press took care of this in its determination to present a story in which a revolution was begun by a rare mould floating through an unsuspecting bacteriologist's window. This was excellent copy. Nor was the Oxford team helped by Florey's initial refusal to talk to reporters.

But, as Florey suspected, Fleming's elevation to hero status also required a certain amount of behind-the-scenes manipulation. St Mary's publicity team was certainly actively involved. The first press announcement of the development of a new wonder drug was despatched to *The Times* in August 1942 by none other than Almroth Wright. His letter was just one part of a broader campaign orchestrated by the head of St Mary's, Lord Moran, who was closely connected to Winston Churchill and several press barons. The very survival of St Mary's depended on charitable contributions, so it was entirely understandable that its senior staff should have exploited Fleming's relatively modest contribution to the development of penicillin to raise the profile of their hospital.

Further, acutely aware that the nation was in dire need of good news in the two years since Dunkirk, the British propaganda machine rolled into action to great effect. Her armies were making painfully slow progress in North Africa, but Britain now had a hero of giant proportions at home. With the help of the Canadian-born press baron, Lord Beaverbrook, and Fleming's own willingness to play the game, he was 'built up' into a national icon. If those involved made a conscious choice, it may also be significant that they went for the native-born Scot rather than the Australian Florey. The only consolation for those to whom the truth has a special status is that Fleming shared the 1945 Nobel Prize for medicine with both Howard Florey and Ernst Chain.

The achievement of greatness is usually categorized under three headings: those who are born to it, those who achieve it, and those who have it thrust upon them. For Alexander Fleming a fourth, compound order seems appropriate: individuals who, having let greatness slip through their fingers, snatch it back from those who have a better claim.

'A DECOY OF SATAN'

Simpson persisted in the use of chloroform for relief of labour pains, against opposition of obstetricians and the clergy. He was appointed one of the queen's physicians for Scotland in 1847 and in 1866 was created a baronet.

'Sir James Young Simpson', *Encyclopaedia Britannica* (2001).

In the early 19th century patients dreaded surgical operations. Anaesthetic was not discovered until 1842, so patients had to endure excruciating pain. In an amputation, the patient would be held down while the surgeon cut through all the soft tissue and bone. The horror of pain forced surgeons to work quickly, often leading to mistakes and a low survival rate. The first successful steps in the conquest of pain were taken by James Simpson.

There was opposition from those who saw chloroform as unnatural and members of the Calvinist Church of Scotland claimed its use was forbidden in the bible.

BBC website, *Medicine Through Time.*

James Young Simpson is one of those luminaries of the past the basis for whose claim to fame becomes a little hazy on close examination. He was certainly not the founding father of modern anaesthetics. In 1799, nearly 50 years before Simpson entered the field, the English chemist Humphrey Davy discovered nitrous oxide (laughing gas) and, having found that it relieved toothache, suggested it would be of help during surgery. Davy's suggestion was not immediately followed up. But during the 1820s, another Englishman, Henry Hill Hickman carried out extensive animal research in pursuit of an effective anaesthetic gas. Alas,

Left: Sir James Young Simpson (1811–70).

Hickman's experiments were only moderately successful and he died, a disappointed man, in 1830 at the age of just 29.

Britain having missed two golden opportunities, the baton then passed to the United States. There much ill luck attended those who first picked it up. In 1844, Horace Wells, a dentist, used nitrous oxide on a patient seemingly successfully. But after the failure of a subsequent demonstration at Harvard University, he was forced to flee Boston in disgrace. In 1846, William Thomas Green Morton, Wells's former dental partner, did manage to give a successful public demonstration, this time using ether. Morton anaesthetized the patient, then a surgeon took over to remove a neck tumour. Nevertheless, Morton's star was only briefly in the ascendant. Claims made by other Americans rapidly soured his triumph. First, a certain Crawford Long insisted that he had been using ether as an anaesthetic from as early as 1842. Then Charles Jackson, a doctor with a strong interest in chemistry who had advised Morton, claimed to have been the brains behind Morton's success.

Although these disputes were to blight the lives of most of those involved, news of what Morton had achieved rapidly spread to Europe. It was only at this point that Simpson entered the frame. As Professor of Midwifery at the University of Edinburgh, he had long been concerned about the extreme pain endured by many women during labour. As a result, on learning what had happened in the United States, in January 1847 Simpson became the first person to use ether in obstetric practice. Yet, Simpson found the strong smell of ether unpleasant. He also noticed that it causes coughing spasms in some patients. So he set about finding an alternative. A friend who was a chemist suggested chloroform, which they first tried on themselves. As it proved at least as effective as ether, and seemed to have none of the side-effects, it had become his preferred anaesthetic by November 1847. Within a month he had used it successfully on more than 50 patients.

Never bashful in getting news of his successes abroad, Simpson's initiative led to the widespread use of chloroform in general surgery as well as in midwifery. Indeed, in 1853, its use in obstetrics received royal approval when John Snow administered it to Queen Victoria during the birth of Prince Leopold. The Queen's appreciation was reflected in Simpson's subsequent baronetcy for services to medicine. He was also honoured internationally and, on his death in 1870, Simpson's family was given the option of a state funeral in Westminster Abbey. They settled for burial in Edinburgh, together with a striking statue on Edinburgh's Princes Street.

There are just a couple of discordant notes to this story. First, Simpson was severely criticized by John Snow for using an anaesthetic at too early a stage during labour. Then, around 1900, it was discovered that chloroform can cause serious liver damage. Ether was therefore re-introduced. And, in time, this, too, was displaced as the modern science of anaesthesia brought about the progressive introduction of the far more effective and much less harmful drugs and gases used today.

But from this brief summary, it is clear that Simpson's was, by any measure, a very distinguished career. Nonetheless, his eligibility for the premier division of scientific greats is distinctly questionable. The best case that we can make for him is that he was a very early user of a technique pioneered elsewhere; the first to utilize anaesthetics in obstetrics; a major popularizer of their use in all branches of surgery; and somebody who introduced a new anaesthetic, which, although seeming to be an improvement, in the long run proved to be less beneficial than the substance it had replaced. There can be no doubt that Simpson was on the side of the angels, but even when a Royal baby is factored in for good measure, there is not enough here to have propelled him to the pinnacle of scientific achievement. So why is he still talked about?

The legend-making ingredient I have so far omitted was central to the story of Simpson's life I was taught when first studying the history of medicine at secondary school. The crux of this account was that Simpson's use of anaesthesia to relieve the excruciating pain of childbirth plunged the medical profession into a bitter and protracted conflict with the Victorian Church. Theologians and churchmen were vehement in their insistence that any attempt to spare women from the pains of childbirth amounted to a direct contravention of God's wishes. For, according to the Bible, after Eve's transgression in the Garden of Eden, God had cursed her saying: 'I will greatly multiply thy sorrow and thy conception; in sorrow thou shalt bring forth children; and thy desire shall be to thy husband, and he shall rule over thee.'

To churchman, therefore, it seemed clear that the pain of childbirth was meant to be endured by all women as the tainted heirs of Eve. By relieving it, Simpson had unwittingly stepped on 'forbidden ground'. Very courageously, given the power of the Victorian Church, he refused to accept the Church's callous teaching. Instead, he threw the full weight of his own humanitarian resolve against religious dogma to ensure that women would gain the right to have childbirth made less agonizing. A savage and prolonged struggle then ensued. And it was only settled when Queen Victoria was persuaded to ignore the objections of the Church by having recourse to chloroform during the birth of Prince Leopold. This royal imprimatur ensured that Simpson and his supporters carried the day. Thereafter the use of anaesthesia in obstetrics became routine and scripture was righteously suborned.

This story provides a fascinating companion piece to the Huxley–Wilberforce legend I told in Chapter 10. Knowledge of the 'religious opposition to anaesthesia' is still widespread and is regularly cited in introductory texts on obstetrics and the history of medicine. It was also one of the most prized examples used by Andrew D. White in his best-selling *A History of the Warfare of Science with Theology in Christendom* (1898). Although published over a century ago, this and books of its ilk have profoundly shaped how we perceive the relationship between science and religion. As White's no-nonsense title makes clear, we are again encountering a worldview that sees a natural antipathy between faith and reason. White characterized himself as one of the victors at the end of a century during which scepticism became acceptable. *A History of the Warfare of Science with Theology* gives his account of the long drawn-out struggle. To him, Simpson's trials and tribulations in overcoming Church resistance to chloroform represent a strategically crucial campaign in a much-broader war. This is how he speaks of the storm whipped up by Simpson's humanitarianism: 'From pulpit after pulpit Simpson's use of chloroform was denounced as impious and contrary to Holy Writ.' He goes on to state that: 'texts were cited abundantly, the ordinary declaration being that to use chloroform was "to avoid one part of the primeval curse on women" '.

Perhaps White thought he was telling nothing but the truth; but what we are actually dealing with here, as with the Huxley–Wilberforce legend, is the political use and abuse of history by an emergent scientific elite. For the claim that the Church in Britain—whether Anglican, Catholic, or Protestant non-conformist—obstructed the introduction of anaesthesia to obstetric wards is little more than a fabrication. Although, on occasion, the Church has demonstrably been prepared to take illiberal positions against scientific advance, such occasions are extremely rare. Instances of institutionalized religion opposing the prevailing opinions and practices of orthodox science have long been exceptional. This is partly because Churches have learned the advantages of accommodation, but more because the interests of science and religion seldom conflict. The case of obstetric anaesthesia is a case in point.

'The religious objection'

The idea that religious bigotry impeded the obstetric use of anaesthesia, and that countless women thereby suffered unnecessary pain, stood unchallenged until the closing years of the twentieth century. This is despite the fact that virtually every reference to the supposed conflict could be traced back to only two sources—John Duns's hagiographic biography of James Young Simpson (1873) and Simpson's own (1847) *Answer to the Religious Objections Advanced against the Employment of Anaesthetic Agents in Midwifery and Surgery*. Sensing that something was amiss, during the 1980s an enterprising historian called A. D. Farr began to look back into the religious, scientific, and popular literature of the 1850s. He wanted to know if this legendary debate had actually occurred. The silence, to use the playwright Sheridan's phrase, assaulted his ears.

Although it is only natural to assume that Simpson's *Answer to the Religious Objections* was a response to a major challenge, Farr searched in vain for evidence of a genuine controversy. This is not to say that no voices were raised in opposition. But it is now apparent that such objectors as there were comprised a tiny, uninfluential minority. The most extensive literature searches have thrown up only three examples. With the first, critical opportunism seems to have been a major factor. In 1847, J. Parke, a Liverpool surgeon, wrote a pamphlet entitled *Reasons for Not Using Chloroform Except in Cases of Extreme Necessity*. Having visited Simpson in Edinburgh in October 1847, Parke had raised some concerns of a technical nature. He made no mention whatsoever of the 'primeval curse' or any other religious argument. It was only after Simpson published his *Answer to the Religious Objections* that Parke brought out his own pamphlet seizing on the new line of attack Simpson had presented him with. As a result, *Reasons for Not Using Chloroform* included the claim that:

> You do not really bless a woman by removing the pains of labour—her true blessing flows from lifting up her heart to God, and asking for humility and strength to bear them. Over and over again, have I seen such faith rewarded, with far more comfort than chloroform could give.

Another British writer, the rather obscure George T. Gream, argued in his *The Misapplication of Anaesthesia in Childbirth* (1849) that because anaesthesia is a form of intoxication, it should be 'esteemed a crime by the laws of God and Man'. Likewise, the American doctor Charles Meigs condemned 'any process, that the physician sets up, to contravene the operation of those natural and physiological forces that the Divinity has ordained us to enjoy or suffer'.

Apart from these examples we have only Simpson's words to rely on. In his *Answer to the Religious Objections*, he wrote:

> Not a few medical men have, I know, joined in this same objection, and have refused to relieve their patients from the agonies of childbirth, on the allegation that they believed that their employment of suitable anaesthetic means for such a purpose would be unscriptural and irreligious.

Later in the same pamphlet, Simpson referred to a 'Dublin man' whose extreme opposition to obstetric anaesthesia Simpson claimed had actually motivated him to write his *Answer*. Finally, in a letter he sent in July 1848 to Protheroe Smith, an obstetrician at St Bartholomew's Hospital in London (and the first obstetrician to use anaesthesia in England), he explained:

> Here, in Edinburgh, I never now meet with any objections on this point, for the religious, like the other forms of opposition to chloroform, have ceased among us. But in Edinburgh matters were very different at first: I found many patients with strong religious scruples on the propriety of the practice. Some consulted their clergymen. One day, on meeting the Rev. Dr H——, he stopped me to say that he was just returning from absolving a patient's conscience on the

subject, for she had taken chloroform during labour, and so avoided suffering, but she had felt unhappy ever since, under the idea that she had done something very wrong and sinful. A few among the clergy themselves, for a time, joined in the cry against the new practice. I have just looked up a letter which a clergymen wrote to a medical friend, in which he declares that chloroform is (I quote his own words) 'a decoy of Satan, apparently offering itself to bless women: but, in the end', he continues 'it will harden society, and rob God of the deep earnest cries which arise in time of trouble for help'.

Intriguingly, however, Simpson offers no support for the idea that only the birth of Prince Leopold silenced public disaffection. According to him, all substantial opposition had ceased a little over a year after he had introduced the technique.

A much more striking aspect of the three documented objections is that they were advanced not by senior clerics but by medical men. Here we have further evidence that the notion of a deep-seated conflict between science and medicine is largely mythical. The middle years of the nineteenth century witnessed a period of widespread evangelical revival throughout the British Isles, and elsewhere, that swept up members of all professions. Even so, despite the strength of mid-century evangelism, Parke, Gream, and Meigs constituted a tiny minority of their profession. There was no general resistance within their fraternity to the use of anaesthesia in midwifery. Nor is there any evidence that the simultaneous introduction of anaesthesia in Europe encountered significant resistance from within the profession.

Were most doctors exceptional in their enlightened attitude? Far from it. It would seem that hardly anyone from any profession—including the Church—objected to Simpson's innovation. Consequently, to create the impression of a significant body of resistance to his use of anaesthesia, Simpson was obliged to construct enemies where none really existed. The 'Dublin man' referred to in his *Answer to the Religious Objections* was later identified as Professor William Montgomery. When Montgomery learned of the way in which he had been portrayed, he vehemently denied having put forward any religious arguments to obstruct the relief of suffering. Having stewed on the matter over Christmas, in a letter of 27 December 1847 Montgomery fulminated to Simpson:

> You say you were induced to write your 'Answer' by being informed that I was publicly advocating these so called 'Religious objections' and that I had denounced you ex cathedra as acting in an unchristian way . . . I never advocated or countenanced either in *public* or in *private* the so called 'Religious objections' to anaesthesia in labour, but invariably rejected that objection.

In a later paper in the prestigious *Dublin Quarterly Journal of Medical Science*, Montgomery announced unambiguously that he attached absolutely 'no value to what are called the "religious objections" to the use of this remedy'. In other words, Simpson's largest crumb of evidence for there being a conflict over the use of anaesthesia bordered on the libellous.

The view from the pulpit

When Simpson spoke of clerical opposition in his *Answer to the Religious Objections*, he referred only to 'a few among the clergy themselves'. This was not then the stuff of edicts, bulls, and decrees with the whole might of the Christian church arrayed against the obstetric use of anaesthetics. In fact, whilst Farr managed to uncover a total of seven references to the religious dimension of anaesthesia in the British and American religious journals of the period, not a single one was critical of Simpson's practice and five of them were emphatically supportive of the obstetric use of ether or chloroform. Most of these articles were also inspired by Simpson's *Answer* and their authors had apparently not even imagined there could be objections before this curious pamphlet was published. Indeed, several years after it had appeared, Simpson himself remarked that he had received 'a variety of written and verbal communications from some of the best theologians and most esteemed clergymen here and elsewhere, and all churches, Presbyterian, Independent, Episcopalian, etc.,—approving of the views which I had taken'. This sounds like mass endorsement rather than entrenched opposition. Unfortunately none of the 'written' communications survive. We do know, however, what a couple of senior churchmen had to say on the subject.

The Reverend Dr Thomas Chalmers was Moderator of the Free Church of Scotland and a former Moderator of its General Assembly, or Kirk. He was an influential religious opinion-former and one of the chief arbiters of religious orthodoxy in Scotland. In 1847, he was asked to contribute an article for the *North British Review* on the theological aspects of anaesthetization. By all accounts, he took some convincing that the request was actually made 'in earnest'. Once reassured that he was not the intended victim of an academic joke, he is reported to have 'thought quietly for a minute or two, and then added, that if some 'small theologians' really did take a strongly negative view of the subject, he would advise . . . "not to heed them"'. Thus we have one of the highest church authorities in Scotland ignorant of a debate that was supposedly raging around him whilst himself taking a very positive line on the use of anaesthesia. We also have a written record of what Oxford University's Reverend Charles Kingsley, most famous as the author of *The Water Babies*, thought on the subject. In addition to his remarkable literary skills, Kingsley was a heavyweight theologian and historian. Sometime during 1852, a member of the aristocracy (whose name we do not know) seems to have raised the religious aspects of obstetric anaesthetics with him. The following reply suggests he thought objections almost laughable:

> The popular superstition that [labour-pain] is the consequence of the fall I cannot but smile at—seeing as it is contradicted by the plain words of the text which is quoted to prove it—'I will greatly *multiply* thy sorrow and thy conception', . . . It being yet a puzzle to me, as a Cambridge man, how the multiplication of 0 can produce a number. $0 \times A$ used to $= 0$, did it not?

The joke may be somewhat laboured, but it makes very clear which side of the argument Kingsley favoured.

Both Chalmers and Kingsley were liberal theologians, so their views may have been untypical. Providing absolute proof that no arch-conservatives took a contrary view is clearly impossible as some objectors may never have put their objections in writing. But the want of written evidence testifies to the quality of Chalmers's intuition and encourages modern historians of science in the belief that for all Arnold White's claims, Simpson's engagement in the alleged war between science and theology was inconsequential, if not entirely imaginary. Seemingly even the most reactionary of bishops saw no merit in entering the lists. James Young Simpson was defending himself against an attack he had largely dreamt up.

So why resort to print?

If only a tiny minority of doctors and a few 'small theologians' were at all bothered about the use of anaesthetic during labour, from where did Simpson derive the impression that he was under siege? To complicate matters, another pamphlet defending the obstetric use of chloroform was published in 1848 entitled *Scriptural Authority for the Mitigation of the Pains of Labour, by Chloroform, and Other Anaesthetic Agents*. Its author was Dr Protheroe Smith, an English disciple of Simpson's, with whom Simpson was in close correspondence. Smith, too, launched what proved to be a phoney war. Having gone to great lengths to prove that the use of chloroform is sanctioned by the Bible, he elicited virtually no reaction at all. By 1848, it is quite clear, any approximation of a debate was long dead. So explaining why Smith rushed into print is even more difficult than accounting for Simpson's *Answer to the Religious Objections*. One thing, however, is apparent. Whatever interests their successors and hagiographers represented, Simpson and Smith were not attempting to advance the cause of positivist science against what they perceived as a stale and anachronistic religion. Neither man shared the aspirations of Thomas Huxley and Joseph Hooker. On the contrary, in understanding why these men raised the question of religious objections it is crucial to appreciate that both men were deeply committed Christians and church-goers.

John Duns, Simpson's biographer, reports that in the 1830s his subject's religious convictions amounted to no more than a 'baptized heathenism'. But during the following decade this all began to change. Simpson's growing spirituality was publicly demonstrated in 1843 when an internal dispute resulted in about a third of the ministers in the non-conformist Church of Scotland leaving to form the Free Church of Scotland. Simpson was one of many laymen who felt sufficiently strongly about the underlying issues to join the exodus. Then, in 1844, his eldest child died in very tragic circumstances and a daughter passed away in infancy 3 years later. These events, and their emotional repercussions, contributed to his intellectual interests broadening out in the following months and years to encompass theology and recondite biblical criticism. When he came to write his *Answers to the Religious Objections*, Simpson was able to refer to several academic biblical commentaries. By then, he had become a distinctly pious man.

For his part, Protheroe Smith has been described as a 'staunch Evangelical Christian'. His 41-page pamphlet contained no fewer than 190 biblical references, and many of these were closely aligned to his particular—dispensationalist—brand of Christian belief. Smith was a genuine and well-informed believer. The same may be said of the obstetrician, Dr John Tricker Conquest, who strongly defended Simpson and Smith's theological reasoning in an 1848 book entitled *Letters to a Mother*. Significantly Conquest was also the author of an 1841 edition of *The Bible with 20,000 Emendations*.

Quite apart from the insight this gives us into the mentalities of those who chose to go into print in defence of Simpson's use of chloroform, it also cuts much of the ground away from Arnold White's notion of an eternal war between science and religion. Here we see three men clearly positioned on the progressive wing of medicine and all with strong religious convictions. At this stage at least, professional boundaries had not been erected between science and theology. Nor, seemingly, had radicals within the profession yet come up with the idea of seeking to elevate themselves by repudiating orthodox religion.

We can feel entirely confident that Simpson and Smith were unexceptional in the nature of their beliefs. Hundreds of other doctors and dozens of obstetricians who felt no need to defend their use of anaesthetics in print would have held similar convictions. Yet, Simpson and Smith differed from their peers and colleagues in one important respect: they were the *pioneers* of the obstetric use of anaesthesia. Attention—favourable and unfavourable—was therefore sharply focused on them.

Speculatively we can suggest several reasons why this may have made all the difference. A psychologist might see projection at work. Because they had such profound religious knowledge, the words from Genesis might have caused both Simpson and Smith deep disquiet. On this reading, both men would have been conducting a public exercise in self-persuasion. A more prosaic possibility is that, as Simpson implied in his July 1848 letter to Smith, the real problem lay with the patients. It may be that to encourage stoicism, generations of ministers had told their female parishioners that they simply had to put up with the pain of childbirth because that was God's will. In a religious age, with this belief being passed on from mother to daughter, it might be that many of the women—not their ministers—needed to have their scruples assuaged. In this sense, Simpson and Smith may have been forearming their professional colleagues with the means of bringing such relief.

A not-incompatible possibility is that Simpson and Smith were attempting what the military term a pre-emptive strike. If church and public had taken exception to what they were doing, their efforts would have been wasted and their reputations destroyed. It may be true that neither obstetrician was able to cite more than a couple of religious complaints lodged by clergymen or laymen. It is also the case that the only reference to the debate that John Duns was able to cite in his section 'communications from patients' was a letter from a lady explaining that had it not been for Simpson's pamphlet, it would never have occurred to her that anyone could object to the obstetric use of chloroform. Nevertheless, because they were taking all the risks attendant upon being pioneers, Simpson and Smith would have been very sensitive to any potential criticism from the religious quarter. Having much to lose they would have been quick to overestimate the possible intensity of religious opposition. Perhaps they also reasoned that their pamphlets, which made very clear the strength of their religious convictions, could do nothing but good for their cause. At a minimum, it would prove that, far from being Godless freethinkers, they took their religion very seriously indeed.

One other (less creditable) motive has also to be considered. In terms of being first overall in the field of anaesthesia, Britain had unquestionably missed the boat. Given that Simpson had been a pioneer of anaesthetics in obstetrics, there may have seemed merit in making this appear as radical a step as possible. On this reading, giving the impression that controversy surrounded their work was a sure means of attracting attention and of presenting the persona of the daring pioneer. So were *Answers to the Religious Objections* and *Scriptural Authority for the Mitigation of the Pains of Labour* in part image-building exercises on the part of two obstetricians in search of immortality? Here the famous Scottish verdict 'Not proven' seems apposite. At the most it could only have been a background consideration. As we have seen, both Simpson and Smith were serious about their religion and it would be rather uncharitable to conclude that the risks of being ostracized by the Church would not have been at the fore-front of their minds.

Science, religion, and myth

In researching this chapter I asked numerous friends and colleagues whether or not they had heard that churchmen had objected to the obstetric use of anaesthesia. Several were familiar with the story. Those who were not replied with striking consistency, 'I hadn't heard of that, but I'm not surprised'. Why is it so easy to elicit this inherent cyncism about the motives of the church? Part of the answer may lie in the modern liberal's angst about the attitude of the Roman Catholic Church, and several less-ancient denominations, towards birth control, homosexuality, single-motherhood, and euthanasia. I suspect, however, that the chief reason is historical and stems from the success of Thomas Huxley, Arnold White, and their equally pugnacious supporters, in promoting their own scientific worldview. Personally I do not doubt that these men were right about the Bible being the work of man. Yet, in presenting religion as anti-science they were playing a rather cynical game in which science was self-consciously cast as the transcendentally rational antithesis of meta-physics and belief.

The most fascinating aspect of Huxley and White's campaign is that one can see the myth of an embattled science in the process of construc-tion. Both men characterize as endemic a conflict that only ever existed in exceptional circumstances. White's *History of the Warfare of Science with Theology* was a landmark on the road to this polarization. 'I propose to present', he opened his tirade, 'an outline of the great, sacred struggle for the liberty of science—a struggle which has lasted for so many centuries, and which yet continues.' Such claims built on pre-existing prejudices concerning the Catholic Inquisition and fundamentalist, Puritan bigotry. In the decades following its publication, with 'witch-hunt' becoming common parlance for aspects of Nazism, communism, and, albeit on a much more modest scale, McCarthyism, it became easy to see science as the one true cause, forever beset by those who feared its revelations.

Yet, when we look again at the examples Arnold White used so successfully to place religion in the dock with the other offenders, we find serious manipulation of the historical record at almost every stage. To uphold the image of a timeless war between faith and reason, White had to suppress huge amounts of pertinent data. The religious beliefs of John Dalton, Michael Faraday, Louis Pasteur, William Thomson (Lord Kelvin), James Clerk Maxwell as well as Simpson and Smith were conveniently omitted. Had they known the anti-religious ends to which their endeavours were to be put, each of these heroes of science would have been appalled. This, however, was not something that troubled—or perhaps even occurred to—White's multitudinous readers. Speaking in 1924, the humanist Edwin Mims insightfully wrote:

> White's *Warfare of Science with Theology* is responsible for much of [the] thinking about religious bigotry and intolerance, and they are ready to join in smiting the Infamous. In other words, college professors are like most human beings in not being able to react to one extreme without going to the other.

As Mims saw, White's science–religion dichotomy was being talked into existence. In the United States, of course, there would be a backlash that is continuing to this day. But in Britain, science and religion did part company. Clerics were largely deprived of their entitlement to speak on substantive matters concerning man's place in nature. This class of person, one that had made major contributions to the advance of scientific knowledge, was rudely cast aside. Centuries in which it was hard to define where science ended and religion began were abruptly terminated in a crudely imperialistic campaign that drove the religious off into reservations whose material barrenness rendered them of little interest to the new scientific professionals. Today, religion largely restricts itself to values, feelings, hopes, and fears, and hard-to-pin-down ideas of a grander purpose in life. But even these reservations are looking less and less sacrosanct to many scientists. To some, neuroscientists and evolutionary theorists seem increasingly like the avaricious gold-diggers who, in defiance of treaty obligations with the native Indians, swarmed over the Black Hills of Dakota in nineteenth-century America.

Yet, blowing the whistle on scientific propaganda is most certainly not to condemn the detachment of science from matters of religious import. Indeed, it is hard to overstate how much science has gained by moving beyond the inflexible 'truths' of the Bible. In addition, we need to recognize that the appeal of Huxley and White's sermons to most scientists had little to do with empire-building. The rhetoric exampled in White's *Warfare of Science with Theology* strikes a chord in the same way as do patriotic hymns and national sporting triumphs. It has served to confer on generations of scientists a vicarious pride and a strong sense of being part of something important. It has allowed them to feel that they are the heirs to a glorious and righteous tradition. As they examine the contents of countless Petri dishes, pore over complex calculations, undertake what is for the most part the grind of scientific research, as well as teach sometimes indifferent students and mark usually unexceptional papers, scientists can cling on to an image of theirs being a noble and chivalric pursuit. Science, they can reflect, is a discipline that transcends petty human conflict and has brought the once-powerful Church to its knees, not in prayer, but in defeat.

For the most part, then, no real malice was intended towards the Church. Thus, in 1925 the American zoologist Winterton C. Curtis explained how 'as a college student in the mid-nineties, I had almost wished I had been born twenty years earlier and had participated in the Thirty Years War [between Darwinians and Christians], when the fighting was really hot'. All saints, it seems, need sinners. But, however we understand the split between science and religion, it is instructive to remember that only a century and half ago James Young Simpson and Protheroe Smith were more worried about the religious implications of what they were doing than was the Church itself.

CONCLUSION TO PART TWO

Sins against history?

I n Part 1 of this book we looked at five case studies in which the central issue was the extent to which the individuals discussed were open to the charge of 'conduct unbecoming of a good scientist'. The unifying theme in Part 2 has been offences committed against the historical record. In this concluding chapter I want to look in more depth at how history came to be so comprehensively re-written; how the approach to history extolled by the Roman historian Pollio was widely disavowed in favour of the myth-making of his rival Livy; whose interests were served; and how those who effected the changes managed to get away with it. We may begin by considering the extent to which some of the individual scientists we've look at were personally responsible for the myths that have grown up around their names.

In the cases of Joseph Lister and Charles Best the evidence now seems overwhelming that they greatly altered the historical record to their own reputational advantage. Of the two, Best appears the greater sinner because of the exceptional lengths he went to in seeking to denigrate the efforts of others. It now seems clear that he was the least significant of the four contributors to the insulin breakthrough, yet he was unable to rest until he had garnered most of the credit. There is considerable poetic justice in his having to recognize in his final years that in pushing this to the degree he did, he came perilously close to destroying his country's claims of primacy in the field of diabetes research. In comparison, Lister's one redeeming feature was a lesser enthusiasm in denigrating the efforts of rivals. Instead, he seems usually to have preferred to wait until they had passed on before quietly attaching their achievements and ideas to his own record.

Were the culpabilities of Lister and Best being judged by a legal tribunal rather than an historian, psychological reports would almost certainly be called for before sentences were passed. Although their actions cannot be condoned, modern psychological science has given us a much better understanding of the constructivist nature of memory. In short, the tribunal would need guidance to gain some insight into the extent to which either, or both, deluded themselves in seeking to delude everybody else. Further enquiries might also extend to gauging the culpabilities of contemporaneous 'powers that be' in supporting or tolerating false versions of events simply because of the lustre they added to institutional and national reputations.

With Thomas Huxley and James Young Simpson the issues are a little less clear cut. Certainly there is something Best-like in Huxley's rewriting of the 1860 Oxford debate. If we put personal pride aside, however, we can see that his overall strategy was as much concerned with upgrading the status of science and the professional scientist as with personal aggrandisement. There is a clear analogy here with a people feeling under an imperative to occupy a territory of their own, brutally displacing the existing occupants to acquire one. Here, too, it is far from uncommon for the successful to be seen by their heirs as heroes, with their darker deeds being swept under the carpet. But to the judicial or historical eye such behaviour does not reflect well on the aggressor. So it seems to me with Huxley. If the religious needed to be driven from the heartlands of science, the only acceptable reason would have been that their science was of an inadequate standard. Vilification by a rival faction may be both expedient and commonplace, but it is not to be admired.

Superficially, it might seem that Simpson is open to identical criticism as he, too, accused men of religion of anti-progressive behaviours on a scale that there is little or no evidence they displayed. But Simpson can be distinguished from Huxley on grounds of motivation. Unlike Huxley, Simpson was a committed Christian and would have been most unlikely to find common cause with him in seeking to drive the religious from the temple of science. As has already been made clear, what actually motivated Simpson in drawing attention to the religious objections to the use of anaesthetics in obstetrics remains uncertain. It may well have reflected a conflict being fought out in his own imagination. It may have been an overreaction to objections raised with him directly by members of the particular congregation to which he belonged. It might have been a strategy for drawing attention to a development that was otherwise not quite as radical as is commonly thought. Whatever the underlying cause, it was not Simpson, but others, who attached his story to the anti-religious bandwagon. He may be open to a charge of reckless publicity seeking; but not one of factional infighting.

There are overlaps here with the forging of Alexander Fleming's reputation by outside parties. Yet, unlike Simpson, Fleming has been accused by some historians of being no more than a competent 'technician', who unjustly garnered the rewards appropriate for a scientific great. The first part of this judgement cannot be allowed to stand because, as was seen in Chapter 12, Fleming's scientific work was of a high quality and his decision to scale down penicillin research during the 1930s was made on entirely credible scientific grounds. But there is a measure of truth in the second accusation. After all, Fleming was apparently happy to acquiesce in the successful attempt to raise him to the status of the lone genius. And to this end he was less than generous to his biochemist colleagues. Nevertheless, a couple of extenuating circumstances might be noted. First, the image of the solitary crusader was in large measure a role created for Fleming by the hospital for which he worked and several favourable news editors. Second, even if he did seek scapegoats for his failure systematically to pursue penicillin work, by the 1940s Fleming could not possibly have expected a public, eager for unblemished heroes, to understand why he hadn't tried to protect an infected mouse with his 'mould-juice'. Whatever his gut feelings, Fleming's only alternative to playing the game was to fall far lower in the public's esteem than he really deserved to go.

If Fleming sinned more by omission than commission, John Snow, Charles Darwin, and Gregor Mendel are almost certainly innocent parties. No compelling evidence has so far emerged that suggests they sought to obfuscate or alter what they had actually done, said, thought, or written. I may have been able to point out, for example, that Darwin erroneously claimed to have developed his theories only after he had accumulated a warehouse full of facts, but this seems to have been no more than a self-delusion common to much of the scientific community. In all three cases, responsibility for the major discrepancy between the legend and the actuality has to be laid at the door of subsequent generations; and this leads us on to the question of why such things happen.

In trying to answer it, Snow is in some ways the easiest case with which to start. Although what epidemiologists do isn't always obvious, it is one of those fields in which the unsympathetic observer is inclined to believe that it is. Thus a typical maestro of hindsight will insist that the value of comparing the health records of intermingled households making use of one or other of two separately sourced water companies is something that should have been obvious from the outset, not a highly imaginative piece of field research. One means of buttressing a profession against this kind of self-serving criticism is to have a well-developed foundation myth showing just how purblind the world was before the new profession emerged. For established epidemiologists seeking to impress on society the importance of their work, or inspire students with what they may be able to achieve, or motivate field workers in the midst of very risky investigations, what better example could be chosen than John Snow? Thanks to a brilliant, trained mind and a willingness to intervene physically when this was called for, Snow succeeded against a cruel and virulent disease where others, locked into ill-founded and untested theories, failed. So perfectly do Snow and his pump handle meet the specification for an ideal foundation myth, we can be certain that the intervention of mere historians has no serious prospect of derailing the legend.

With Darwin I think another factor can be detected. Evolutionary theory has always had powerful enemies who attack it not on the grounds of being obvious, but on grounds of its being a nonsense. Whether it be the Duke of Argyll writing in direct response to Darwin's publications or the modern critique known as intelligent design, these opponents have a supreme confidence in their ability almost to laugh the theory of natural selection out of existence. Superficially they have a point. Natural selection may not seem counter-intuitive to the dizzying extent achieved by quantum mechanics, but initially it is very hard to accept that the near infinite variation and subtlety of nature could have resulted from no more than marginal advantage interacting with selective pressures over immense time scales. That what are now known as neo-Darwinists can so robustly defend their position against such attacks is a measure of the extent to which they have developed and refined Darwin's original ideas. The incorporation of modern genetics into evolutionary theory has produced what seems to be an impregnable redoubt from which those at the forefront have become famous for taking the fire to the enemy.

For some, however, there seems to be a deep-seated desire for a founding father very much less equivocal than the one history actually supplied. Not for them a Darwin who never fully gave up the idea that environmental pressures can directly call forth structural adaptations during the lifetime of individual organisms. Nor one who clung on to the Lamarckian belief that the lifetime experiences of a parent can have direct implications for the characteristics with which its young are born. Instead, the old master's voluminous and wide-ranging opus is used selectively to present a man as confident in the explanatory powers of natural selection alone as his modern heirs. Again, it is unlikely that the work of historians will significantly change the situation. But if truth rather than hero-worship became the primary consideration, those interested in Darwin's own ideas could do worse than heed a variant of the advice senior civil servants used to give their juniors when first encountering Winston Churchill, 'Remember, he is a Victorian!'. In Darwin's case the appropriate caution would be, 'Remember, to a considerable degree he's a pre-Darwinian'.

With Gregor Mendel we have an exceptional advantage in probing what led to the creation of the myth. A book exists that is a literary parallel of the famous fossil *Archaeopteryx*: the one is a crossover from dinosaur to bird, the other a crossover from Mendelian reality to Mendelian myth. Robert H. Lock, a Cambridge biologist, published *Recent Progress in the Study of Variation, Heredity and Evolution* in 1906 with a second edition coming out in 1909. The work was among the earliest to introduce Mendel's ideas to a general readership. The first significant finding is that if we look in its index, we do not find Mendel's laws, but Mendel's law. If we then turn to the first page number given, which is midway through the book, we find:

> We are now in a position to state the important proposition known as Mendel's law, which is to the following effect: The gametes of a heterozygote bear the pure parental allelomorphs completely separated from one another, and the numerical distribution of the separate allelomorphs in the gametes is such that all possible combinations of them are present in approximately equal numbers.

Note that this conforms precisely to what careful re-examinations of the text suggest was the extent of Mendel's understanding. In short, as I have already argued, Mendel was not proposing a general rule but identifying something he considered peculiar to hybrids. Even the in-text definition Lock gives for 'allelomorphs' (the original word for alleles) speaks not of genes but of characters.

Lock's index reference carries another page number towards the end of the book. If we turn to that we find not a restatement of the above, but 'the law based by Professor Correns upon the conclusion which [Mendel's] paper contains'. The essence of this reformulation is that:

> the cells of zygotic organisms—organisms, that is to say, which have arisen by the process of sexual reproduction—contain a double complement of hereditary qualities. Such cells may contain A and A, a and a, or A and a. The forms AA and aa are described as homozygotes, the form Aa as heterozygote.

Within the space of about 100 pages, therefore, Mendel's ideas have been extended from what he considered the special case of hybrids to encompass all life forms engendered by sexual reproduction. A man who almost certainly died still puzzled by the patterns of reproduction he found, is credited with having fully comprehended the ubiquity of the gene-pair. Having made this attribution, Lock feels able to give as his avowed opinion that Mendel's brief paper is the 'most important contribution of its size which has ever been made to biological science'. Here Lock lavishes exceptionally high praise on Mendel and, in doing so, seems to be oblivious to the major discrepancy between what he acknowledges Mendel actually said and the ideas he later attributes to him.

This case is particularly interesting because there is no direct national pride at stake. After all, what nationalistic interest could a Briton have in exaggerating the achievements of a deceased Abbot of Brno? Nor could Lock have been in any doubt as to who amongst his contemporaries were making the breakthroughs he was partially re-assigning to Mendel. His book includes numerous references to Carl Correns, Erich V. Tschermak, and Hugo de Vries, each of whom independently claimed to have rediscovered Mendel's paper after having previously covered much the same experimental ground and (in some cases) reached rather more profound conclusions.

Lock's behaviour can be explained in several possible ways, some of which we have already encountered in Chapter 7. First, faced with three strong rival contenders for the genetic laurels, Lock and his fellow professionals were being asked to make their equivalent of the Judgement of Paris. Reaching back in time to Father Mendel obviated the need to make an invidious choice. Second, with many biologists setting themselves up in opposition to the concepts of gradual evolution and natural selection, the need for a single, rival icon may have seemed pressing. Third, the dead are more malleable than the living. Anointing Correns, Tschermak, or de Vries as the founding father risked empowering the chosen one to lead the emergent discipline as he saw fit. A long-dead Abbot presented no such risk. Fourth, the Mendel story could be made to fit the standard hero model so easily. Once he was credited with having cracked the genetic underpinnings of sexual reproduction, the unenthusiastic way in which his ideas were received fitted perfectly with the mandatory period a true hero has to spend in the wilderness. Who better to have as your titular leader than one whose ideas were so far in advance of their time that his contemporaries found them incomprehensible?

Whatever the relative weightings of each of these factors in Mendel's case, the last is probably of the widest relevance. It may also have a concealed psychological benefit. For whatever reasons—cultural and/or evolutionary—we humans seem to be obsessed by hierarchy. Crucially, in any given group, the overall status of the group leader determines and limits the status opportunities available to subordinates. Looked at from this perspective, what we have seen to be a not uncommon practice of exaggerating the achievements of chosen icons may well serve to elevate the status of those responsible. The New Testament makes clear that St Paul spoke with pride in saying to his jailers, '*Civis Romanus sum*' ('I am a citizen of Rome'). Perhaps Lock, having made his contribution to the enhancement of Mendel's reputation, was saying with equal satisfaction, 'I am a Mendelian'.

But where does this leave those more interested in getting as close to the truth as is now possible? I think the answer should be fairly clear. We must look at what our heroes of science really said at the time rather than taking them at either their own or their disciples' later valuations. If we really want to understand the past as it happened, we must seek out those whose work shows clear signs of their having followed the approach summed up by the Renaissance humanist in the phrase '*Ad fontes*!'. The effective researcher goes back to the original 'springs' or sources, rather than accepting the traditions that have become accreted to, and obscured, the truth.

The examples of Lister, Fleming, and Best further counsel us to be extremely cynical about autobiographical accounts of discovery. They strongly suggest that only the most self-effacing of individuals—such as John J. Macleod and James Collip—can resist exaggerating their personal contributions to great scientific breakthroughs. Each of the preceding chapters also spells out the need for scepticism when dealing with much of the existing literature. Many biographers seem to have been written to glorify idols, ancestors, disciplines, or even the nations in which the scientists lived and worked. All too often, knowingly or unknowingly, the biographer has left the historical truth trailing in the wake of legend.

If nothing else, I hope to have reminded the reader of, or alerted them to, phrases which should immediately trigger suspicion: 'The man before his time', the 'wilderness years', 'the conservative opposition' are the most common. All of these concepts converge on the idea of the lone genius. But, as we have seen, the genius of conventional imagery—as someone who sees far beyond the petty intellectual concerns of his or her period—is a great rarity. Most of the 'great' scientists of the past 200 years (such as Lister, Mendel, Darwin, Snow, and Fleming) were doing things that many of their scientific contemporaries could immediately under-stand and, in some cases, were already doing themselves. Furthermore, some of these 'Gods among men' were ploughing theoretical furrows that have only a tangential relation to modern ways of thinking. To summar-ize, perhaps without exception, each member of our Pantheon of scientific heroes has benefited from one or more of the four basic misrepresenta-tions elaborated below:

- The cogency of their evidence in their own time has been exaggerated.
- Their distinctiveness has been overplayed and contemporaries with similar ideas have been unceremoniously sidelined.
- The incremental steps to a new theory, requiring the separate contribu-tions of many individuals over many decades, have been ignored or downplayed. And
- Past theories with a vague affinity with modern ideas have been torn out of context and force-fitted to our modern understanding.

With these points in mind, we should unhesitatingly question any historical account that fails to show an awareness of these pitfalls. In particular, any work that tries to claim that a single man or woman was responsible for a massive breakthrough that eluded all of their predecessors and contemporaries, deserves to be treated with the utmost circumspection.

We need also to be sensitized to the dangers of evaluating the quality of past scientists against inappropriate yardsticks. To reach a balanced judgement we must exclude much of what we know today in evaluating a dead scientist's skills as a scientist. As earlier chapters showed, having an idea later proved right may well owe much more to good luck than good science. A fair evaluation relies on looking at each case in context. Such an approach enables us to see that for entirely rational reasons, our predecessors have believed in many things that we now know to be utterly misconceived. A century and a half ago it made perfect sense for Darwin to think that growth and reproduction are tightly linked: that there is a close correspondence between the growth of hair and nails and the production of new offspring. And, without knowledge of DNA and the isolation of germ cells in the reproductive organs, it was eminently sensible to believe that new individuals are made from buds splitting off from their parents' body cells.

It was similarly entirely reasonable for Mendel to conclude that hybrids were a special reproductive case. Only when later scientists looked at his results with the eyes of those who had seen paired chromosomes and the genetic basis of fruit-fly traits could 'Mendelian' genetics be empirically derived. Even Joseph Lister, who retro-engineered his past, had good reasons for doubting the validity of modern germ theory when it first began making headway in the 1880s. Hindsight is a wonderful thing but, if not checked, it will lead the historian wonderfully astray.

What the cases we have looked at strongly suggest is that true pioneers are exceedingly rare. So we have to scale down our definition of 'great' and allow ourselves to admire lesser but much more real achievements. At the same time, we need to shift our perception of the history of science from an Olympic relay race, stretching across the ages, to a more sophisticated—though admittedly less inspiring—notion of gradual, cumulative effort in which only the rewards are not widely distributed.

Had this more realistic perception of science been embraced during the 1940s, perhaps Fleming and Best would have felt a lesser compunction to claim an importance neither of them actually had. It would also serve to reduce hostility to the idea that, for example, neither Darwin nor Mendel got everything right. Indeed, unless we modify our view of normal science, it is hard to square the fact that Darwin and Mendel were exceptionally able men with the realities of what they did. The real fault, I believe, lies not with them, but with our over-inflated expectations.

Aside from being more accurate, this view of science as incremental but most definitely progressive is surely far more congenial to practising scientists themselves. Does it really buttress the psyche of the individual scientist to be given the impression that, say, more than 99 per cent of researchers are menial plodders whose chief historical importance is to provide the less than 1 per cent of geniuses with ignorant opposition and then rapturous applause? Science's great strength is the willingness of its practitioners to build on the achievements of others; or, in disproving them, to observe things about the world that had previously gone unnoticed. As this suggests, no scientist is an island. Instead, most are members of tightly linked networks of highly specialized researchers who depend on others' expertise, advice, and experience for their own ideas to make any progress at all. As well as involving conflict and controversy, good science is necessarily collaborative and co-operative.

Nor by any stretch of the imagination can modern science be said to be unsuccessful. More has been accomplished by modern scientists than by their precursors in all other periods in the history of humanity combined. Science is more efficient, more rigorous, more streamlined, and better funded today than Darwin, Pasteur, or Fleming could ever have imagined. This in itself may help to answer the complaint of many living scientists that there are no modern Newtons or Einsteins. Some have argued that the savagery of peer review, through which all prospective articles have to pass, crushes out radical new ideas. As I made clear in discussing Robert Millikan and Louis Pasteur, this is a view with which I have considerable sympathy. But we have also to reflect on the possibility that there is now much less scope for Newtons or Einsteins to emerge.

Epoch-making discoveries require the opening up of provinces of Nature that have been left largely untouched since the birth of modern science. But the sheer scale of the modern scientific enterprise must reduce the possibilities of finding such untrammelled territories. It would be foolish to think that all the great discoveries have already been made—this is unlikely to be the case—but there is a major difference between the scientific worlds of the late nineteenth and twenty-first centuries. Now scientific fields are sufficiently packed with highly able men and women that progress is made ever more evenly. There can be no doubt that the overall productivity of scientific effort has been hugely increased, but there is much less scope for one individual to shine 'like a moon amongst stars'.

In closing I would like to re-emphasize a point that may have become somewhat obscured along the way: that the aim of this book has not been to denigrate science. Rather, my chief targets have been an overly simplistic reading of what science is all about and the strong tendency to romanticize its past achievements. There are probably ardent defenders of the dignity of science, however, who will take extreme exception to any form of demythologizing in the history of science. Revisionist scholars are seen as pandering to base instincts and plying their trade with an unseemly relish. Perhaps the chief irony of such a stance lies in the failure to recognize that in so condemning the new history of science, critics implicitly insist on historians turning a blind eye to their source materials in a manner that would be considered disgracefully unprofessional within the laboratory setting. A professional historian can no more be content to glamorize than a scientist to embellish or invent. He or she has an unavoidable professional obligation to study the past in as scientific a manner as can be reasonably managed. Indeed, the essential correctness of this approach should be more obvious to the stalwart defender of science than to almost any other category of human being. In short, getting close to the truth really matters.

This is a point that could hardly be more simply or eloquently expressed than in this extract from the section of Robert Graves's *I, Claudius* I discussed in my Introduction. The background is Pollio having mocked Livy for his cavalier attitude to the truth:

> Livy came slowly towards us. 'A joke is a joke, Pollio, and I can take it in good part. But there's also a serious matter in question and that is, the proper writing of history. It may be that I have made mistakes.

What historian is free of them? I have not, at least, told deliberate falsehoods: you'll not accuse me of that. Any legendary episode from early historical writings which bears on my theme of the ancient greatness of Rome I gladly incorporate in the story: though it may not be true in factual detail, it is true in spirit. If I come across two versions of the same episode I choose the one nearest my theme, and you won't find me grubbing around Etruscan cemeteries in search of any third account which may flatly contradict both—what good would that do?'

'It would serve the cause of truth,' said Pollio gently. 'Wouldn't that be something?'

Chapter 1 (pp. 15–31)

This chapter is based on the research and interpretations of an American historian, the late Gerald L. Geison. His *The Private Science of Louis Pasteur* (Princeton University Press, 1995) helped establish the private laboratory notebook as the essential source for the modern historian of science. Geison, who died in July 2001, was Professor of History at Princeton University. This account is also indebted to research undertaken with Geison by another historian, John Farley, and published under the title 'Science, politics and spontaneous generation in nineteenth-century France: the Pasteur–Pouchet debate', in the *Bulletin for the History of Medicine* (vol. 48, pp. 161–98, 1974). Another account of this controversy is contained in a book written by two sociologists of science, Harry Collins and Trevor Pinch, entitled *The Golem: What You Should Know about Science* (Cambridge University Press, 1998). The term 'experimenter's regress' was introduced by these scholars. Finally, for a general history of the debate about spontaneous generation, see John Farley's *The Spontaneous Generation Controversy from Descartes to Oparin* (Johns Hopkins University Press, Baltimore, 1977).

Chapter 2 (pp. 33–46)

The principal source for this account is an essay by Harvard Professor of Physics and Professor of History of Science, Gerald L. Holton. Holton's essay 'Subelectrons, presuppositions, and the Millikan–Ehrenhaft dispute' was published in his book *The Scientific Imagination: Case Studies* (Cambridge University Press, 1978; pp. 155–98). He stresses the necessity of jettisoning some experimental data and the need sometimes to go beyond the available evidence to explore potentially important ideas. The other main source is an article written by Alan D. Franklin, 'Millikan's published and unpublished data on oil drops', published in *Historical Studies in the Physical Sciences* (vol. 11, 187–201, 1981). Franklin adjusts some aspects of Holton's argument but upholds the general view that Millikan was less than scrupulously honest when publishing his raw data. For more wide-ranging analyses of the nature, dynamics, and difficulties of experiment-ation see Harry Collins and Trevor Pinch's *The Golem: What you Should Know about Science* (Cambridge University Press, 1998) and Peter Galison's *How Experiments End* (University of Chicago Press, 1987).

Chapter 3 (pp. 49–63)

I based this chapter on the account of John Earman and Clark Glymour entitled 'Relativity and eclipses: the British eclipse expeditions of 1919 and their predecessors', published in *Historical Studies in the Physical Sciences* (vol. 11, 49–85, 1980). They conclude that although Eddington clearly did doctor his results, he did so out of a strongly emotive conviction that relativity is a 'beautiful and profound theory'. Further information and most of the quotations used in this chapter were drawn from Harry Collins and Trevor Pinch's *The Golem: What you Should Know about Science* (Cambridge University Press, 1998). For more on the role of trust in science, see Steven Shapin's *A Social History of Truth: Civility and Science in Seventeenth-Century England* (University of Chicago Press, 1994) and the book he co-authored with Simon Schaffer, *Leviathan and the Air Pump: Hobbes, Boyle and the Experimental Life* (Princeton University Press, 1985). Finally the best modern biography of Sir Isaac Newton is Richard Westfall's *The Life of Isaac Newton* (Cambridge University Press, 1993).

Chapter 4 (pp. 65–76)

This chapter is based largely on the research of two professors of manage-ment and business administration, Charles D. Wrege and Amedo G. Perroni. Their detective work, motivated by the conviction that histori-ans must present the real evidence whether or not we like what emerges, was first published under the title 'Taylor's pig-tale: a historical analysis of F. W. Taylor's pig-iron experiments' in the journal *Work Study and Management Services* (vol. 9, pp. 564–9, 1974). Charles D. Wrege and Ronald G. Greenwood have recently published a collaborative biography of Taylor, entitled *The Father of Scientific Management: Myth and Reality* (Business One Irwin, Homewood, 1991). Most of the quotations includ-ed in this chapter are drawn from F. W. Taylor's *Scientific Management* (Harper and Booth, New York, 1947).

Chapter 5 (pp. 79–98)

The key source for this chapter is Alex Carey's 'The Hawthorne Studies: a radical criticism', published in the *American Sociological Review* (vol. 32, pp. 403–16, 1967). Also interesting is Dana Bramel and Ronald Friend's 'Hawthorne, the myth of the docile worker, and class bias in psychology', published in *American Psychologist* (vol. 36, pp. 867–78, 1981). These pieces have prompted several other reappraisals that the writers of textbooks and teachers of management courses are only now coming to appreciate. But a close reading of Roethlisberger and Dickson's *Management and the Worker* gives enough ammunition for plenty more critiques.

Chapter 6 (pp. 115–31)

The principal source for this chapter is a paper that appeared in *The Lancet* (vol. 356, pp. 64–8, 2000) entitled 'Map-making and myth-making in Broad Street: the London cholera epidemic, 1854', by Howard Brody, Michael Russell Rip, Peter Vinten-Johansen, Nigel Paneth, and Stephen Rachman, all of whom are academics at Michigan State University. Also useful has been the follow-up correspondence in volume 356 of *The Lancet* by Jan P. Vandenbroucke, of Leiden University Medical Centre, and David Morens of the National Institutes of Health, Bethesda. Vandenbroucke has written several papers on the Snow myth; for example, his 'Who made John Snow a hero?' in the *American Journal of Epidemiology* (vol. 133, pp. 967–73, 1991). The early history of epidemiology is examined in L. G. Stevenson's 'Putting disease on the map: the early use of spot maps in the study of yellow fever', published in the *Journal of the History of Medicine* (vol. 20, pp. 226–61, 1965). I have also gleaned valuable information from a website dealing with all things related to John Snow and cholera located at www.ph.ucla.edu/epi/snow.html. For those interested in the broader history of attempts to understand, control, and cure cholera, the best text is Charles Rosenberg's *The Cholera Years: the United States in 1832, 1849, and 1866* (University of Chicago Press, 1987).

Chapter 7 (pp. 133–58)

Key sources for this chapter were Robert Olby's *The Origins of Mendelism* (University of Chicago Press, 1985), Augustin Brannigan's *The Social Basis of Scientific Discoveries* (Cambridge University Press, 1981), and L. A.

Callender's 'Gregor Mendel—an opponent of Descent with Modification', published in the journal *History of Science* (vol. 26, pp. 41–75, 1988). See also Garland Allen's biography *Thomas Hunt Morgan: The Man and His Science* (Princeton University Press, 1981), Peter Bowler's *The Mendelian Revolution: The Emergence of Hereditarian Concepts in Modern Science and Society* (Athlone, London, 1989), and Loren Eiseley's *Darwin's Century: Evolution and the Men Who Discovered It* (Doubleday Anchor Books, New York, 1958).

Chapter 8 (pp. 161–75)

This chapter is based largely on the research of three historians of medicine and is contained in two separate articles. First, Christopher Lawrence and Richard Dixey's 'Practising on principle: Joseph Lister and the germ theories of disease' in Lawrence's book *Medical Theory, Surgical Practice: Studies in the History of Surgery* (Routledge, London, 1992; pp. 153–215). Second, Lindsay Granshaw's ' "Upon this principle I have based a practice": the development and reception of antisepsis in Britain, 1867–90', in the book *Medical Innovations in Historical Perspective* edited by John V. Pickstone (Macmillan, Basingstoke, 1992; pp. 17–46). A good biography of Lister is Richard Fisher's *Joseph Lister, 1827–1912* (Stein and Day, New York, 1977). For up-to-date surveys of the rise of modern medicine see Roy Porter's *The Greatest Benefit to Mankind* (HarperCollins, London, 1999) and Charles Rosenberg's *The Care of Strangers: The Rise of America's Hospital System* (Johns Hopkins University Press, Baltimore, 1987).

Chapter 9 (pp. 177–203)

This chapter is based on the work of numerous historians working over several decades. Rather than list dozens of individual books and articles, it is perhaps best if I simply mention a few of the more influential and readable texts. For the reader wanting to find out more about Darwin, his life and work, I recommend two biographies: Adrian Desmond and James Moore's *Darwin* (Penguin, London, 1991) and Janet Browne's *Voyaging* (Cape, London, 1995). In addition, Peter Bowler's *Evolution: The History of an Idea* (University of California Press, Berkeley, 1989) contains a full but concise analysis of the modern history of evolutionary theory. Adrian Desmond's *The Politics of Evolution: Morphology, Medicine, and Reform in*

Radical London (University of Chicago Press, 1989) is an important resource on pre-Darwinian evolutionist thought and Robert J. Richard's *Darwin and the Emergence of Evolutionary Theories of Mind and Behaviour* (University of Chicago Press, 1987) and his *The Meaning of Evolution* (University of Chicago Press, 1993) cover in detail most of the individuals mentioned in this chapter. Stephen Jay Gould's *Ever Since Darwin* (Norton, New York, 1979) offers many additional insights.

Chapter 10 (pp. 205–21)

This chapter is based on several essays, foremost among them are J. Vernon Jensen's 'Return to the Wilberforce–Huxley debate', in the *British Journal of the History of Science* (vol. 21, pp. 161–79, 1988) and J. R. Lucas's 'Wilberforce and Huxley: a legendary encounter', in the *Historical Journal* (vol. 22, pp. 313–30, 1979). Adrian Desmond's new two-volume biography *Huxley* (Michael Joseph, London, 1994 and 1997) looks in detail at this fascinating and acerbic man. For general accounts of the relationship between science and religion in Britain and the United States, see: Frank M. Turner's *Between Science and Religion: The Reaction to Scientific Naturalism in Late Victorian England* (Yale University Press, 1974); John Hedley Brooke's *Science and Religion: Some Historical Perspectives* (Cambridge University Press, 1991); and Edward J. Larson's *Summer for the Gods: The Scopes Trial and America's Continuing Debate over Science and Religion* (Basic Books, New York, 1997). Finally, Peter Bowler's *The Eclipse of Darwinism: Anti-Darwinian Evolution Theories in the Decades around 1900* (Johns Hopkins University Press, Baltimore, 1983) describes the demise of Darwinian thought during the late nineteenth century.

Chapter 11 (pp. 223–45)

As mentioned in the text, this chapter is drawn almost entirely from Michael Bliss's book *The Discovery of Insulin* (University of Chicago Press, 1982) and his article entitled 'Rewriting medical history: Charles Best and the Banting and Best myth' published in the *Journal of the History of Medicine and Allied Sciences* (vol. 48, pp. 253–74, 1993). Also useful is Ian Murray's 'Paulesco and the isolation of insulin', again in the *Journal of the History of Medicine and Allied Sciences* (vol. 26, pp. 150–7, 1971).

Chapter 12 (pp. 247–67)

This chapter is drawn from Gwyn Macfarlane's biographical study, *Alexander Fleming: The Man and the Myth* (Chatto & Windus, London, 1984). Also valuable is Ronald Hare's *The Birth of Penicillin* (Allen & Unwin, London, 1970). F. W. E. Diggins's articles, 'The true history of the discovery of penicillin, with refutation of the misinformation in the literature', in the *British Journal of Biomedical Science* (vol. 56, pp. 83–93, 1999) and 'The discovery of penicillin; so many get it wrong', in *The Biologist* (vol. 47, pp. 115–19, 2000), usefully criticize the attempts of some to reduce Fleming's status to that of a 'third-rate' scientist.

Chapter 13 (pp. 269–83)

This chapter is based on the scholarship of A. D. Farr, published under the title 'Religious opposition to obstetric anaesthesia: a myth?', in the journal *Annals of Science* (vol. 40, pp. 159–77, 1983). A. D. White's interpretation has been severely criticized in three books: John Hedley Brooke's *Science and Religion: Some Historical Perspectives* (Cambridge University Press, 1991); Edward J. Larson's *Summer for the Gods: The Scopes Trial and America's Continuing Debate over Science and Religion* (Basic Books, New York, 1997); and Peter Bowler's *Reconciling Science and Religion: The Debate in Early-Twentieth-Century Britain* (University of Chicago Press, 2001).